BEACHWALKER
SEA LIFE OF THE WEST COAST

**STEFANI
HEWLETT
PAINE**

Douglas & McIntyre
Vancouver/Toronto

Douglas & McIntyre Ltd.
1615 Venables Street
Vancouver, British Columbia
V5L 2H1

CANADIAN CATALOGUING IN PUBLICATION DATA

Paine, Stefani Hewlett, 1946–
    Beachwalker

    First ed. has title: Sea life of the Pacific
Northwest
    Includes index.
    ISBN 1–55054–016–5
    1. Marine fauna—Northwest Coast of North
America.   I. Title.   II. Title: Sea life of the Pacific Northwest.
QH95.3.P34 1992   591.92'53   C92–091081–5

PHOTO CREDITS

Cover: Blood star (*Henricia leviscula*), courtesy Vancouver Public Aquarium.

All colour photographs are by Finn Larsen except for Plate 8, lined chiton, by Ernie Cooper, and Plate 9, Steller's sea lions, by Roy Tanami.

Black and white photographs
R. Bayer: 58, left.
Margaret Butschler: 30; 133, top; 140, top right; 147.
Ernie Cooper: 91.
Pierre Dow: 12; 15, bottom; 35; 46; 50; 53; 57; 58, top left; 63; 72, top middle; 80,
middle; 99, top and bottom; 102, middle; 106; 127; 137; 138; 139, top middle.
John Ford: 159.
Gilbey Hewlett: 161.
Peter Hulbert: 146.
Finn Larsen: 9; 24; 26; 33; 34; 36, bottom; 38; 56, top left; 58, lower left; 62, top and bottom; 64; 66, top; 67; 72, top left; 73; 94; 98, bottom; 101, top right; 111; 112; 114, bottom; 115, bottom; 118; 121; 124, top and bottom; 125, top and bottom; 126; 130; 131; 132; 133, bottom; 134; 135, top;
138, top left; 139, bottom left; 149; 155; 162; 165.
Ron Long: 72, top right; 76; 81; 96; 107; 140, top left.
Gar Lunney: 90.
Roy Tanami: 151; 154.
Vancouver Public Aquarium: 36, top; 44, bottom; 52; 72, bottom; 89; 98, top; 115, top; 135; 136; 140, bottom; 142; 152; 166.
Vancouver Sun: 129; 144.
Jim Willoughby: 31; 139, bottom right.

Those photographs not otherwise credited are by the author.

Editing by Nancy Flight
Design and typesetting by Eric Ansley & Associates
Printed and bound in Canada by D. W. Friesen

CONTENTS

PREFACE

Through my experience working with the public at the Vancouver Public Aquarium, camping, and talking with boaters, divers, fishermen and families out enjoying the beach at low tide, I have become aware of the tremendous interest people have in what they see. So often I have been asked, "Where can I get a good book on—?"

Many excellent publications are available on various aspects of marine life of the West Coast. However, many of these publications assume that the reader has a background in biology and are therefore too technical for the layperson to use with ease.

This book attempts to help the lay reader identify marine life of the West Coast, to provide some information about the natural history of the organism and to stimulate questions in the mind of the observer. How big does the organism grow? What does it eat? Why does it live here? How does it reproduce? Who are its enemies? Can it be eaten? Does it bite?

This book includes a representation of commonly encountered marine life from sea plants to whales, with the exception of sea birds, which are well documented in other books. *Beachwalker* does not pretend to be a definitive work on the marine organisms of the West Coast; it does attempt to be accurate and readable for the audience it was written for—the beachwalker, the fisherman, the boater, the camper, the diver and anyone who has seen the teeming life of a Pacific tide pool and marvelled at it.

<u>*A C K N O W L E D G E M E N T S*</u>

Much of the material for this book was drawn from a previous work, *Sea Life of the Pacific Northwest.* I am indebted to Pierre Dow for his great contribution to the first book and to him and Mona Dow for support on this book. Thanks are again due to Greg Davies for his wonderful illustrations and to Gil Hewlett for his participation in the first book.

Beachwalker had lots of help, and to these people I extend sincere thanks: to Vicki Haller, forever enthusiastic, willing and efficient, to Rob Sanders for proposing that I do this book, to my editor, Nancy Flight, for patience, precision and good humour, and to Eric Ansley for wonderful design. Very special thanks to Roy Tanami for his critical eye and pictures and Finn Larsen for enthusiasm, participation and beautiful photographs. Thanks go to others who contributed photographs: Ron Long, John Ford, Jim Willoughby, Margaret Butschler, Ernie Cooper, Peter Hulbert and Gil Hewlett. Finally, there are some people who simply through their warmth, friendship or support help create a positive environment where effort can flourish, and so I thank James and Isabelle Graham, my dear friend Wendy Bradley, my parents, to whom I dedicate this book, my son, Brook, and most of all my husband, Michael Paine.

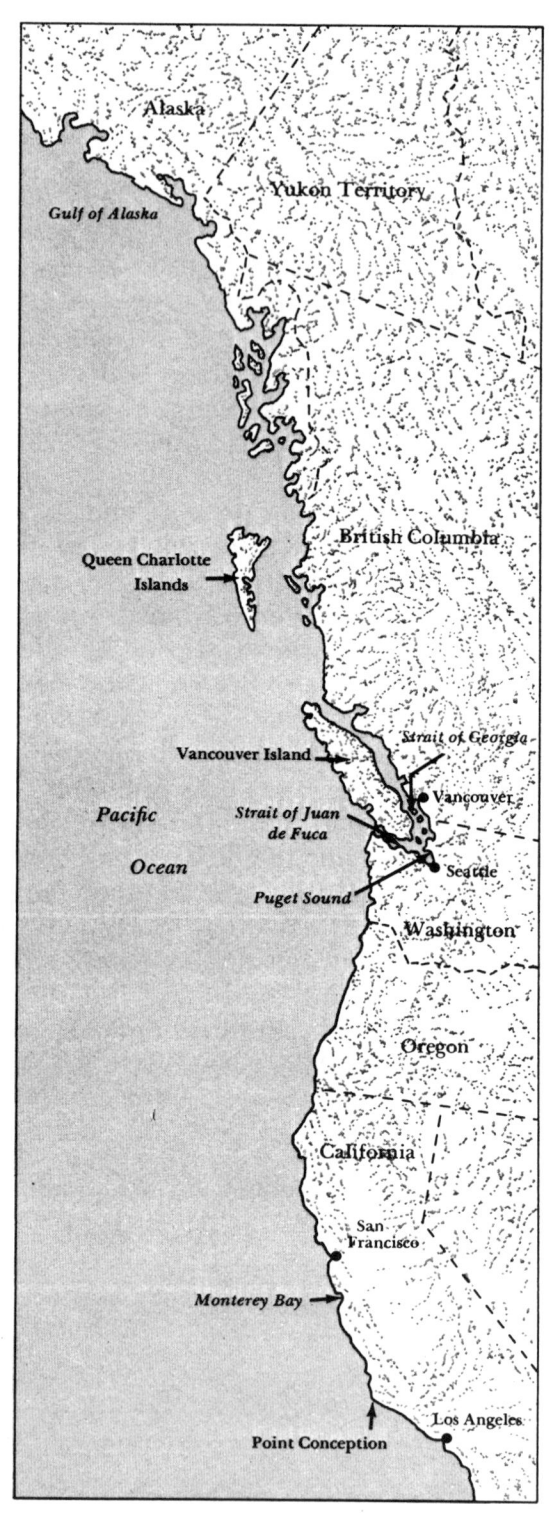

The West Coast region.

INTRODUCTION

Everybody can enjoy a marine life adventure. It is free, no special equipment is needed, no particular skills or talents are required, and the opportunity is available year round.

There is, however, one small but critical limitation—the tide. The tide is "in" or the tide is "out." High tide—when the tide is "in"—is a good time to have a snack, wash the car or write letters. When the tide is high, everything is covered up. The animals are still there eating, mating, battling for territories and generally doing the things they do to earn a living. Unfortunately, you, the beachwalker, are excluded. But wait, the ever-moving tide will go out, or fall, exposing a fascinating world to the landlubber. It is at low tide that you must get yourself down to the beach to turn over rocks, look under ledges and lift seaweed to see what is hiding there.

There are three points to remember. First, look in your local newspaper for the day's tides or invest in a tide table. Any low tide is better than high tide, but certain times of the year have lower tides than others, and the lower the tide, the more exciting and varied the marine life. Second, time your beachwalks to begin about an hour before the extreme low tide. This will give you time to follow the tide out to its lowest point instead of being driven up the beach by incoming water. Third, select a beach with a rich potential for life. On land, how many field mice, spiders or tree frogs would you find on a sand dune? None. The reason is that there is nowhere for the animals to hide from their enemies and nothing for them to eat. It is not so different in the marine environment. A lovely, smooth sandy beach is terrific for swimming and building sand castles, but it is not the best place for many animals when the tide is out. Most marine animals need to hide from hungry shore birds and get out of the drying rays of the sun.

Understanding why the animals must hide will help you find them. They are there, under rocks, between boulders and under the protective blades of kelp. Gently lift and peek. Better yet, look for a tide pool, the temporary home of many small fish and even an occasional octopus. Avoid areas near river mouths, since marine animals don't like fresh water.

Many plants and animals described on the following pages are referred to as "intertidal." This means that the organism is known to occur in the shore area that is left exposed when the tide goes out. An organism that is referred to as "subtidal" has no tolerance for exposure and only occurs in areas that are always under water. Chances are you won't see it on a beach.

HOW TIDES WORK

Ours is a water planet and as such is perhaps unique in the solar system. Except in the polar regions, the Earth's average surface temperature falls within the exceedingly narrow range between 0° and 100° C (between 32° and 212° F), where water remains liquid. At lower temperatures, water becomes solid ice; at higher ones, a gas.

Over 70 per cent of the Earth's surface is covered with water. If all the Earth's irregularities were smoothed out both above and below the water, there would be no land at all—the ocean would cover the entire globe to a depth of 3660 metres (12,008 feet). The tallest peak on land, Mount Everest, could be sunk without a trace in the ocean's greatest abyss, the 10 863-metre (35,640-foot)-deep Marianas Trench in the western Pacific.

It is no wonder, then, that the gravitational pull of the sun and moon have an observable effect on such an enormous mass of liquid. The phenomenon is known to us all as the twice-daily rise and fall of the tides. As the moon swings around the Earth every twenty-four hours, a bulge of water appears on the side of the Earth facing the moon. At the same time a bulge of equal size forms on the opposite side. Because the sun is so very far away, it has about half the pulling effect of the moon. However, it is still able to reinforce, or offset, the moon's pull according to its relative position. For example, when the sun and moon are in line—as they are during the full moon—they act together, producing unusually high and low tides, known as spring tides. During the summer along the West Coast, spring tides occur during the day. During the winter, spring tides occur at night. When the moon, sun and Earth are at right angles to each other, as in the moon's first and third quarters, the pull of the sun and moon cancel each other, producing tides of low amplitude known as neap tides.

Tides are also influenced by the shape of the ocean basins and the enclosed land masses. Islands occurring at the centre of their tidal basins experience little difference between tides. For example, the difference between high tide and low tide on the Island of Tahiti averages 30.5 centimetres (1 foot). Tides near the rim of a tidal basin have greater differences. On the West Coast, for example, the greatest tidal difference averages 4.6 metres (15 feet).

THE WEST COAST REGION

From the beachwalker's perspective, one great quality of the West Coast is its abundance of life. As the tide goes out along the West Coast from northern California to Alaska, sea birds, mink, raccoons, crows, eagles and shore birds move onto the exposed shore. They are there to take advantage of a bountiful supply of food in the form of clams, mussels, oysters, worms, crabs and fish. As the water returns, there will be little evidence that a harvest took place, so abundant is the life on many a West Coast shore.

The reason for the great abundance of living organisms on the West Coast is food—food for plants in the form of nutrients in the water and food for animals in the form of plants and other animals. Like terrestrial plants, all marine plants, from the tiny, single-celled floating plants to the giant kelps, absorb dissolved nutrients and with sunlight make food in a process called photosynthesis. The waters of the West Coast are especially rich in nutrients because of a phenomenon known as seasonal upwelling, which occurs between February and July. The seasonal upwelling is a combination of wind, tides, currents and geography that stirs up nutrients from the depths of the ocean beyond the narrow continental shelf and brings them to the surface. There, sunlight enables photo-synthesis to take place, providing an abundance of food for other organisms.

Another feature of the West Coast is the extraordinary uniformity of water temperature, which averages 14° to 15° C (57° to 59° F), from northern Baja California to Alaska. This is the result of a current known as the West Wind Drift moving in the Pacific Ocean from west to east. On reaching the West Coast of North America, the current splits to the north and south. Upwelling cold water mixes with the coastal currents off California, resulting in water temperatures that are uniform enough to enable many species of plants and animals to live throughout the length of the cool-temperate West Coast region. Along the coast of British Columbia and Washington there also exists a mixing of southern species that find the northern limits of their range in this area and of northern Alaskan species that find the southern limits of their range here.

The West Coast also includes a diversity of habitat, and plants and animals vary according to the habitat. Hardy species such as rockweed or shore crabs, which can withstand the rigours of exposure during low tide, are found in the intertidal region. Some species, such as the moon snail, thrive in sandy areas, whereas the majority of marine organisms are adapted to a rocky habitat that provides a solid place of attachment or offers protection. Wave shock must also be taken into account. Animals such as the surf anemone are adapted to withstand the full force of unbridled waves, whereas more delicate organisms would be crushed and torn by only one great wave.

Thus, these plants and animals live where they do because they are adapted to a particular environment. The environment includes the water quality (the degree of salinity and oxygen content), the availability of food, the opportunity to reproduce successfully and the relationship that organisms have with other plants and animals sharing the environment.

No plant or animal can exist as a thing unto itself; all are interrelated and therefore interdependent. If an environment remains stable, the organisms within it coexist in a balanced state. When the environment is changed through natural events or conditions, such as volcanic eruptions or altered currents, or through human changes such as sewage disposal or break-waters, the balance is upset and the plants and animals must readjust their lifestyles—some go and some stay, according to their capacity to adapt.

SCIENTIFIC AND COMMON NAMES

For many people, including many students of biology, the whole business of naming creatures is a colossal headache. There is no problem, really, in naming a particular plant or animal by any name that one thinks is appropriate. There is no law to say that a dandelion cannot be called a whisker ball or anything else. Anyone is able to exercise individual creativity and inspiration in naming plants and animals, just as we do in naming our children or cats, dogs and canaries. But do not expect to be able to communicate with others about the organism. From family to family, community to community, country to country, the common or vernacular names applied to living organisms change.

Many years ago, starting with the Swedish biologist Linnaeus, scientists began to use a standardized system of naming all living organisms, both plant and animal. In most cases two Latinized names are given. The first refers to genus (a taxonomic group), and the second to species. For example, *Felis domesticus* is a domestic cat, *Felis concolor* is the cougar, and *Felis nigripes* is the African black-footed cat. *Felis* indicates the close relationship of the three cats; *domesticus*, *concolor* and *nigripes* distinguish each species as being separate and unique.

Scientific names are used to avoid the confusion that may occur with widely used common names. For example, many North American species were given their common names by European pioneers. When a fish that looks and behaves much like the Atlantic salmon was seen on the Pacific coast, it was named "salmon" and is still known by that name today. However, this fish is not the same as an Atlantic salmon. The former is known as *Salmo salar*, and the latter are really five different species of the genus *Oncorhynchus*.

Scientific names are subject to stringent rules and

come under the authority of an international congress. In contrast, even widely accepted common names may be changed for all manner of reasons. A case in point is the longjaw rockfish (*Sebastes alutus*). A number of years ago, a commercial fishery was formed to market the species as frozen fish sticks. These were sold under the name longjaw rockfish, and sales were poor. Marketing analysts established that poor sales were not due to appearance or palatability, so they advised that the fish be renamed Pacific ocean perch. Now Pacific ocean perch (*Sebastes alutus*), alias longjaw rockfish, sells well, proving that although a rose smells as sweet by any other name, the consumer likes some names more than others.

TO REMEMBER WHILE YOU ARE AT THE BEACH

Enjoy yourself at the beach, but take care of it. Consider yourself a guest of the marine life of the West Coast. Be patient, turn over rocks, sit quietly by a tide pool and observe through your own shadow the community below busy with the business of survival.

Remember that carelessness or thoughtlessness may mean needless death. If you lift or turn a rock, put it back the way you found it; otherwise, some small creature may fry or freeze in a few short minutes. Fill back clam holes, and never disturb eggs of any kind. By all means enjoy the sea's bounty, but take only what can be eaten. Abide by size and catch limits set down by fish and wildlife agencies; they have been invoked for your own future enjoyment, to ensure an abundance of animals for everyone.

If you are not a diver, visit your local aquarium and see, in dryness and comfort, another dimension of marine fauna.

CHAPTER 1

Marine Plants

Marine plants are easy—they don't run and they don't bite. They include the tiny phytoplankton, algae and sea grasses.

There is one type of phytoplankton, called the dino-flagellates, that have both plant and animal characteristics. These tiny organisms are responsible for two amazing phenomena. One is red tide, and the other is commonly referred to as phosphorescence.

The tiny sparkles seen in a calm, black sea, made by an outboard engine, by a dipping paddle or simply by stirring the water with a long stick at night, are from the dinoflagellates. When some kinds of these minute, free-swimming organisms are disturbed or moved, a chemical reaction takes place in their tissues. The reaction produces light and is seen as twinkling sparks. Many other marine animals produce biological light, or "bio-luminescence," but generally only as a glowing dot or spot. It is the millions of tiny swimming organisms all "bioluminescing" at once that is so spectacular. Thus, "phosphorescence" is really a misnomer.

PHYTOPLANKTON: TINY, TINY FLOATING PLANTS

Marine organisms are generally described as plankton, nekton and benthos. Plankton are all those small, drifting organisms, both plant and animal, that have only feeble powers of locomotion and are carried helplessly at the mercy of currents and tides. Nekton are strong-swimming animals such as squid, fishes and whales, whose movements are powerful enough to make them independent of water movements. Benthos are all those bottom-living organisms, such as clams, starfish and sponges, that crawl over the sea bottom, burrow into it, are sedentary in habitat or remain fixed to one spot.

The sea is filled with plankton (from the Greek *plank-tos*, meaning "drifting"). Some of this plankton is animal organisms, or zooplankton, including the larvae or eggs of many species, such as starfish, clams and anemones. A huge portion of plankton is plant material and is called phytoplankton, or phytoplankters. These phytoplankters are so small that dozens would cover the head of a pin. It has been estimated that 1575 metric tons of microscopic vegetable matter per square metre (4000 tons per square mile) is produced annually in the English Channel.

To understand the marine community, it is essential to understand the nature and function of phytoplankton. Because phytoplankters are plants, they typically contain a green pigment called chlorophyll, which they use, with the aid of sunlight, to convert nutrients dissolved in the sea into organic matter. This process is known as photosynthesis. Since the energy derived from sunlight is essential to this process, the phytoplankton must live within the top layers of ocean water, where enough sunlight can penetrate.

Photosynthesis creates oxygen. Phytoplankton growing in huge ocean meadows produces oxygen, releases it to the air and constantly renews the oxygen supply being used by other organisms, including humans. Scientists believe that if the oceans were to become polluted to the point where the phytoplankton was destroyed, land-dwelling plants could not produce enough oxygen through their photosynthesis to sustain life.

All animals eat either plants or other animals, which have, in turn, eaten plants. In the sea a salmon will perhaps eat a herring, but what does the herring eat? The herring eats the zooplankton, in the form of tiny shrimps and larvae. In turn, the zooplankton eat the phytoplankton, and the phytoplankton produce their own food. This is a very simple example of a food chain. In any food chain, plants form the foundation.

Besides being producers of food, the phytoplankton are essential recyclers of matter. Animal and vegetable wastes are broken down in the sea by bacteria, creating a kind of fertilizer. Upwelling currents and turbulence bring these nutrients to the water's surface, where they can be used by the microscopic plants and so again become part of the food chain.

The phenomenon known as red tide occurs when conditions enable particular species of phytoplankton (*Gonyaulax catanella* on the Pacific coast) to reproduce and become so dense as to discolour the water with a red, brown or yellow stain. Concentrations can reach up to 9 million per litre (2 million per gallon). This generally occurs in areas of upwelling (when bottom water moves to the surface), where vast amounts of nutrients are being brought to the ocean's surface, and where high enough surface temperatures occur. Animals such as oysters and clams, which feed by filtering phytoplankton and other small organisms from the water, tend to concentrate a toxin produced by the phytoplankton in their bodies. When a human being eats these oysters or clams, the cumulative effect of the toxin affects the central nervous system, resulting in great pain and discomfort.

A red tide is more likely to occur during the summer months and is a temporary phenomenon. As the phytoplankton bloom and multiply, they soon use up the nutrients they need, and they cease to flourish.

ALGAE

The toothpaste you used to brush your teeth this morning, the last dish of ice cream you ate and even the mayonnaise in your lunch sandwich more than likely contained some by-product of algae. More commonly known as seaweeds, algae are found in all oceans of the world and are incredibly diverse in size, shape and habitat.

Algae differ vastly from land plants. Land plants have established systems that are much more involved than those of algae. They have roots to absorb water and min-

Sea sac (*Halosaccion glandiforme*) is common and abundant at the mid-tide level. Each individual hollow sac is filled with water and may grow from 10 to 25 centimetres (4 to 10 inches) in height. Colour may vary from yellow-brown to olive to reddish-purple.

erals from the soil, stems that conduct these materials, and leaves that manufacture food (by photosynthesis, using chlorophyll, sunlight and the minerals) to support the entire plant. Algae have no such sophisticated system. Photosynthesis in the algae takes place over the entire surface of the plant. Algae have no roots; a "holdfast," which may look like a root, acts simply as an anchor. Algae may have a stemlike stipe, but it performs no transport function. Unlike land plants, algae do not reproduce from seed. Many algae, in fact, bear no resemblance at all to our usual concept of plants. Some look like spongy, bulbous masses of encrusting growths on rocks. Others are masses of fine threads, drifting freely with the waves.

Many algae are cosmopolitan in their distribution, whereas others are specific to generalized climatic zones (Arctic, Temperate, Subtropical and Tropical) and particular conditions within those zones. Vertical distribution is dependent on tide, depth, competition and sunlight. Along the West Coast, the turbidity or cloudiness of the water and the angle of the sun's rays do not allow enough light to penetrate much beyond 30 metres (98 feet) for algae to grow. Farther south, in the tropics, where there is more direct sun and less turbidity, algae is abundant at four times this depth.

Algae differ markedly in their ability to withstand the effects of exposure: drying, freezing and temperature change. Those with the greatest tolerance are found in the intertidal zone, the shore area that is regularly covered with sea water and left exposed as the tide swells and recedes. Some, such as the stubby 2-centimetre (¾-inch)-high *Prasiola*, are so hardy that they exist above the high-tide mark in the splash zone. Some species live in the subtidal zone and are always covered with water.

Yet even in the subtidal zone attached algae will be found close to shore. As the land falls away, the ocean floor becomes deeper and darker and therefore unsuitable for the manufacture of food by photosynthesis. The horizontal distribution of algae is dependent on salinity, degree of exposure and nature of the sea floor. Shore areas near fresh-water outfalls from rivers such as the Fraser River in southern British Columbia are relatively impoverished in marine vegetation. That is because the sea water is diluted with fresh water, increasing turbidity and reducing salinity.

Exposure is important. It takes very sturdy algae, like the large kelps, to withstand the crashing surf of open coasts. Algae of the open coast are generally firm, with strong holdfasts and thick stipes, yet elastic enough to give in the unharnessed seas. Less hardy species favour protected inshore areas like the rocky shores on the east side of Vancouver Island. Here, protection, plus high salinity, creates an ideal environment for luxuriant forests of algae.

Even more critical is the quality of the substrate, or bottom. Sand and loose pebbles offer no security for attachment or anchorage, whereas large, firm rocks are ideal. Many species of algae attach themselves to other attached algae, though they are not parasitic.

Algae are grouped into divisions according to colour: green (Chlorophyta), brown (Phaeophyta), red (Rhodophyta) and blue-green (Cyanaphyta). The green pigment chlorophyll is present in all algae but is masked by other pigments in all but the greens. Green algae more commonly occur in fresh water and in damp soil. They are far less prevalent in the sea. Commonly seen is the beautiful emerald-green sea lettuce (*Ulva lactuca*), which looks very much like its name. Brown algae are the most strictly marine of all seaweeds, with only three small fresh-water forms. Characteristic of exposed shores, brown algae dominate the vegetation of coastal waters in volume, numbers and size, though not in number of species. Most obvious and well known of these are the kelps like bull or ribbon kelp (*Nereocystis*). In Europe the term "kelp" is applied to the burnt ash of seaweeds, whereas in North America "kelp" refers to the large laminar algae such as bull kelp and giant kelp (*Macrocystis*).

The great kelp beds provide a vital service to the marine community, hosting dozens of "roomers" along their blades and stipes. The 30-metre (98-foot)-long *Macrocystis* is the high-rise of downtown marine land, offering an offshore home, in well-aerated water, away from overcrowding on the beach. For boaters, the kelp beds mark dangerous underwater shoals and indicate the direction and speed of currents.

The large bulb of bull or ribbon kelp and the bulb-like structures on other kelps and algae act as floats, ensuring that the plant will lie near the water's surface and obtain enough sunlight. One such alga has been

Sea lettuce (*Ulva lactuca*) is a common species of the upper intertidal zone. It ranges from Chile to the Bering Sea. This species is pleasant tasting—quite edible—but chewy.

Giant kelp (*Macrocystis intergrifolia*) has individual fronds along its stipe, each supported by its own small float. The plant may grow to 30 metres (100 feet) in length and is perennial, living from four to seven years. Since the species occurs in areas of high salinity close to the open sea, yet not directly exposed to heavy surf, beds of giant kelp are an important habitat for many species of young fishes. This species is harvested commercially for algal derivatives.

Marine Plants   13

Rockweed (*Fucus sp.*) is a very common seaweed of the middle and lower intertidal zone. When the plant is sexually mature, the tips become swollen. Rockweed grows to 50 centimetres (20 inches) and is lime or yellowish-green in colour. Many intertidal animals find shelter from sun and wind under rockweed. Gently lift a few clumps and see what's hiding there.

Sargassum weed (*Sargassum muticum*) occurs in the lower intertidal and upper subtidal from British Columbia to Oregon. The long—to 2 metres (6½ feet)—and profusely branched plants provide shelter for many juvenile fish. This species was accidentally introduced from Japan.

called the oyster thief *(Colpomeniasi nuosa)*. This balloon-like plant attaches its holdfast to the shells of oysters, inflates its bulb with gases during low tide and carries the oyster with it as it floats away at high tide.

Another very common brown alga is rockweed *(Fucus)*, also known as popping wrack, a many-branched, flattened alga. When the plant is reproducing, its tips become swollen. This species grows to 30 centimetres (12 inches) in length. Free-floating or attached sargassum or Japanese weed *(Sargassum muticum)* is infamous among anglers but serves as valuable nursery space for many young fishes. This brown alga was introduced to the West Coast along with the Japanese oyster and has flourished ever since. Sargassum weed is remarkably similar to an indigenous alga, the woody chain bladder *(Cystoseira germinata)*, which can be distinguished from the introduced plant by its pointed bladders or air sacs. Those of sargassum are always rounded.

Although there are many species of algae, they do not occur in great numbers. Of the multitude of red species, few are found in the cool, temperate and sub-arctic regions of the West Coast, yet these tend to be proportionately larger than their southern cousins. An easily recognized red alga, is the red rock crust *(Litho-thamnium)*. Looking much like a splotch of red paint, 2.5 to 10 centimetres (1 to 4 inches) in diameter, it is a tiny encrustation and can vary considerably in colour from whitish-pink to deep purple. Another, dulse, or red kale *(Rhodymenia palmata)*, is a dull red, irregularly divided, broad-blade plant, having a texture like thin rubber. This alga is widely used as a food or medicine. Dulse is said to make a tasty relish and can be eaten raw or chewed like gum.

Blue-green algae are inconspicuous in the marine environment, since they are largely microscopic and seen generally as dark, slimy masses.

WHERE TO LOOK FOR ALGAE

When looking for algae, you should venture out at or just before low tide, seeking rocky areas away from fresh-water outflows. Surge channels between rocks are often ideal and will reward the seeker with an Ali Baba's cave of algae and intertidal creatures. Since many algae are annuals, in that the blades and stipes grow anew in the spring, whereas the holdfast is perennial, a fresher, less wave-worn specimen will be found in early summer than in late fall.

Algae provide excellent subjects for collection, preservation and study, as they neither bite nor flee from the collector. Algae specimens may be preserved in 3 to 4 per cent commercial formalin (neutralized with borax) in sea water and kept in a darkened area to preserve their colour. Some of the very delicate varieties lend themselves well to drying and pressing. Partial drying in the sun, to the consistency of leather, then further slow drying between paper, will usually yield a most attractive design. Special herbarium papers and blotters are available for those who wish to take the extra effort to make a more professional mount.

If you are not interested in collecting algae, you may wish to experiment with food dishes using algae. The Japanese use over twenty varieties in their culinary art. Although not all West Coast species may be palatable, none are known to be poisonous. It remains a matter of taste. The nutritional value of algae lies in their mineral and vitamin content. Their vitamin B (thiamin) content compares favorably with that of many fruits and vegetables. However, since most of the algae's complex carbohydrates cannot be digested by humans, algae are considered a poor source of energy.

SEA GRASSES

Two genera of seed plants, as opposed to algae, occur abundantly in West Coast waters. These are the eelgrasses, *Phyllospadix* and *Zostera*. Although superficially they appear grasslike, they are not related to true grasses. These perennial, rooted marine plants are thought to have originated in fresh water and subsequently adapted to a marine environment. These plants bloom under water; the threadlike pollen is carried by water rather than air or insects, as in most land plants. *Zostera*, or true eelgrass, is dull pale green, and has longer, more delicate blades than *Phyllospadix*. It is found growing over mud and sand bottoms in sheltered areas and is a favourite food of sea birds and many other marine animals. *Phyllospadix*, also known as false eelgrass or surf grass, is easily distinguished from *Zostera* by its bright emerald-green color, shorter length (up to 91 centimetres, or 3 feet) and wiry blade. This is a plant of open, exposed waters.

Sea palm (*Postelsia palmaeformis*) occurs on upper tidal, exposed coasts and in heavy surf. It grows to 60 centimetres (24 inches) in height. Stripe and holdfast are perennial, whereas blades are renewed each year. A strong, flexible stipe keeps the plant from being ripped from the rocks by heavy surf.

Surf grass (*Phyllospadix sp.*) is found on rocky coasts exposed to the full force of ocean waves and is the favoured habitat of many invertebrates, which find a degree of protection within the grass bed. Worms, isopods, many snails and fish such as the snailfish can be found on the grass blades or near the roots. When in flower, male and female blooms are found on separate plants. Beachwalkers can use a small dipnet to find all kinds of small animals living among and along the blades of sea grasses.

CHAPTER 2

Sponges

You will not find the type of sponge you use in the bath on the beach. That kind of sponge does not live on the West Coast, but many others do. Sponges don't do very much, and the kinds you find in the intertidal zone are typically rather shapeless. Some even have a distinctly unpleasant odour.

To identify a sponge, watch for something that looks as if a thick liquid like pancake batter had been spilled on or under a rock. The spill can be any shape or colour. Stroke the surface of the stuff. If it feels slick and smooth, it is probably some kind of algae. If it feels a little like felt or extremely fine sandpaper, it is probably a sponge. Now get a little closer and look very, very closely to see if there are small, slightly raised holes scattered over the surface. If so, congratulations—you have met a sponge!

"To throw in the sponge," an expression commonly used to admit defeat, had its roots in boxing. At the end of the match, the sponge used to mop the fighters between rounds was thrown to the centre of the ring as a token of defeat.

"Sponge cake," "bath sponge," "to sponge off someone" and "to sponge up liquid" are all well-used words and phrases in the English language. They owe their origin to the lowliest of animals, the living sponge.

Sponges are the most primitive of all many-celled animals. Only the single-celled protozoans trail them in development. It is estimated that 4500 species of sponge are living today, all of which are marine except one fresh-water family. The fossil record reveals that sponges were alive as far back as 600 million years ago, a half-billion years at least before even the most primitive human appeared on Earth. During this great length of time the sponges have changed little, carrying out a most basic life history of living, reproducing and dying.

The sponge couldn't be simpler; it is essentially a loose aggregate of cells. The cells are not organized into definite tissues, much less organs, and one wonders why they really bother to stay together. Sponges have no muscle, no nerves, no sensory cells to tell them if it is day or night—not even a mouth.

All sponges are based on essentially the same basic body plan; a vaselike structure attached at one end to something solid, like a rock, a shell or the sea bottom,

Cross-section of sponge wall illustrating direction of water flow

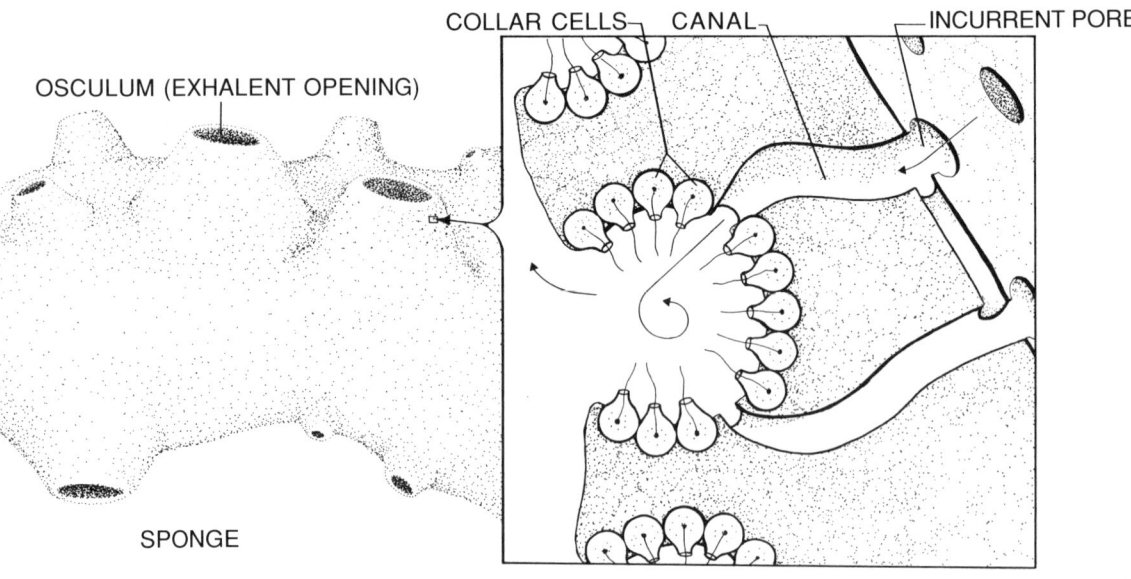

COLLAR CELLS CANAL INCURRENT PORE

OSCULUM (EXHALENT OPENING)

SPONGE

and open at the other end. One layer of platelike cells covers the exterior of the vase. Another layer of cells, called collar cells, lines the interior of the vase. A cellular secretion known as mesoglea is sandwiched between the two layers of cells. The collar cells are like small cups with tiny whiplike tails, known as flagella, extending from their centres into the open interior cavity of the sponge. As the flagella beat, water is drawn into the sponge cavity through the many minute pores, called incurrent pores, that penetrate the sponge's body wall. These pores are too small to be clearly seen. The water exits through the larger opening at the sponge's summit. The water exit is known as the osculum and is generally large enough to be seen with the naked eye.

Although they are still based on the "vase" theme, many sponges have numerous body cavities and oscula, forming an elaborate system of canals. Encrusting sponges, for example, have multiple oscula that look like small volcanoes scattered over the sponge's surface. This kind of sponge may bear little visual resemblance to the vase structure previously described, but it is functionally the same.

Since the sponge lacks a mouth and digestive tract, it is difficult to imagine how its body cells receive nourishment. Minute food particles carried in the water are brought into the sponge cavity, where they become entrapped and engulfed by the individual collar cells. Only particles small enough to fit inside a single cell can be eaten. Digested nutrients are distributed to non-feeding cells by bodies in the mesoglea known as amoebocytes. Waste products are simply released from the individual cells and carried away in the constantly circulating water currents.

Amoebocytes not only deliver the groceries but also produce eggs or sperm for reproduction and secrete a very basic skeleton. The skeleton takes the form of small slivers of glass, calcium or fibre, depending on the species of sponge. *Spicules*, as the slivers are called, give firmness to the sponge and keep it from collapsing on itself and so closing the canals and cavities needed for water circulation.

Because sponge species differ greatly in shape, size and colour, the size, shape and composition of the spicules are often the only definitive means of identifying many sponges. For example, the cloud, or trumpet,

Bread crumb sponge (*Halichondria panicea*). Lumps of this orange, yellow or green sponge can be found on rocks, on kelp stipes, in mussel beds and in similar areas in the mid and low intertidal zone. Despite its often unpleasant odour, it is eaten by the slime star and some nudibranchs. The life span of a colony is thought to be two to three years.

Sponges   19

sponge *(Aphrocallistes vastus)* can be a few centimetres long in fingerlike projections, or it can consist of huge masses as tall as a person and shaped like a trumpet or a large, soft cloud.

The thin encrusting sheets of sponge bear little resemblance to the huge, globular masses of some sponges or the delicate fronds of others. However, some species, particularly those that live in still waters, do grow to a consistent and recognizable pattern. Among them are such fancifully named sponges as lyre, peacock's tail and Neptune's goblet. The tennis ball sponge *(Craniella spinosa)* consistently grows into a fuzzy version of its namesake.

Sponges reproduce both sexually and asexually. Sponges are not male or female but are hermaphroditic, producing both eggs and sperm. The gametes are released at different times to allow for cross-fertilization, meeting by chance as they float in the sea. After fertilization, the larva floats about for a long time before settling to the ocean bottom, never to move again as it grows into an adult sponge.

Like most lower animals, sponges possess remarkable powers of regeneration, or the ability to regrow lost parts. Commercial sponge growers take advantage of this phenomenon by "planting" pieces of cut-up sponge, much as gardeners plant potato slices to produce many potato plants from one tuber. Sponges are also able to regroup their cells if divided. Some species can be strained through fine mesh bolting silk, and given time, the separated sponge cells will reorganize themselves into a complete sponge. No other animal is capable of such total recovery from such extreme mutilation.

Because sponges are totally devoid of movement (save the beating of the flagella), they cannot walk, crawl or drift away from a deteriorating environment. Instead, some will simply disintegrate to a small blob of cells and wait out the situation, regenerating again in better times.

Belying their innocuous appearance, sponges are not entirely inoffensive. The boring sponges attack the shells of many mollusks (shellfish), chemically eating away at the shell and causing it to disintegrate. *Cliona celata*, for example, is seen as small yellow patches on the shells of some scallops. Below the shell is the mass of honeycombed sponge. Boring sponges attack non-living mollusk shells as well and are thought to be substantial contributors to the natural breakdown of shell material littering the ocean floor.

Few animals prey on sponges. The uninviting texture of sponge tissue, due to its spicules, is very likely a deterrent, as is the offensive odour of some species, such as the foul sponge *(Lissodendoryx noxiosa)*, so named for its foul smell. However, some sea slugs and snails persevere and eat sponges anyway. A beautiful little sea slug, *Rostanga pulchra*, has taken on the brilliant red of the encrusting sponge it feeds on. This encrusting red sponge *(Ophlitaspongia pennata)* is seen as red or orange patches on rocks of the lower intertidal level.

From the point of view of evolution, the sponge is a dead end. It was an experiment of nature that worked (it still exists), but it did not give rise to a more complex animal. Somewhere near the main stem of the evolutionary tree, sponges deviated, and for this reason they are often termed Parazoa, meaning "beside the animals," as opposed to the other multicelled animals known as Metazoa.

In some ways sponges could be considered more interesting dead than alive. Venus's flower basket, for example, is the glass skeleton of a deep-water tropical sponge and is frequently displayed in museums. The elastic fibre skeletons of many sponges have had innumerable practical uses since historic times. According to Greek mythology, Glaucus of Anthedon was a sponge diver. Ancient writings tell of Greek soldiers going off to war with sponge padding in their helmets and leg armour. Wives left behind no doubt used sponges to bathe themselves and their children. In times past, sponges were used as cups for drinking; on the cross, Christ was offered a sponge soaked in vinegar.

Most commercially usable species of the "natural" sponge occur only in tropical waters. The live sponges are fished using trawls, poles or divers. They are then cleaned on the sponge boats or taken to shallow ponds and left there until the soft tissues rot, after which the fibre skeletons are squeezed clean in sea water and left to dry. The resulting sponge is the protein skeleton of the once-living animal with all cellular tissue removed. No sponges having this particular kind of skeleton are found on the West Coast.

Today natural sponge is very expensive and has largely been replaced by synthetic sponges of rubber or plastic. It is worthwhile reflecting that the inspiration for these convenient synthetic sponges came from the skeleton of one of nature's dullest creatures.

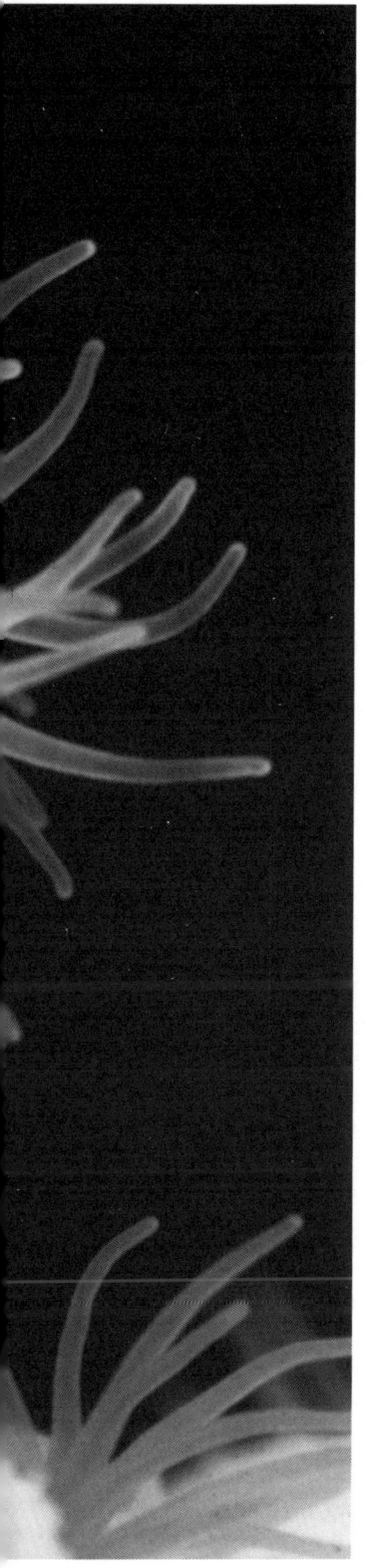

CHAPTER 3

Cnidarians and Comb Jellies

Jellyfish, corals, sea anemones and their relatives are collectively known as cnidarians, or coelenterates. Grouping these organisms together may not appear very logical. But the body plan is basically the same in all of them. Think of them as different models of the same plan: one for swimming, one for sitting and one for many individuals living together with some shared structures. Think of a condominium where the housing units are attached to each other and share corridors, but each unit is still self-contained.

Comb jellies are little globes of jelly that look like small jellyfish. In fact, they are completely unrelated to any cnidarian, but in evolutionary terms they follow the cnidarians.

ABOUT CNIDARIANS

It is not surprising that for many years cnidarians were considered neither plants nor animals but a connecting link between the two kingdoms. These tentacled creatures, often lavishly frilled and beautifully coloured, are strongly reminiscent of springtime flowers. Yet the cnidarians are carnivorous animals, albeit primitive in structure.

There appears to be little similarity between a tiny coral polyp and a floating jellyfish, yet the animals are closely related. They are based on the same simple body plan of a hollow sac—closed at one end, open and fringed with tentacles at the other. Imagine a cup with tentacles surrounding its rim. When the cup is securely attached by its base to some firm surface, tentacles and opening directed upwards, it is known as a polyp. When the cup is free floating and is suspended upside down in the water, with opening and tentacles on the underside, it is known as a medusa, or jellyfish. The jellyfish structure has been named medusa for the snaky tresses of that mythological maiden the Gorgon Medusa.

Many of the cnidarians have both an attached polyp form and a free-swimming medusa form in their complete life cycles. These two distinct forms occur in alternate generations, so the offspring do not look at all like their parents but are identical to their grandparents. The polyp reproduces a medusa, or jellyfish form, by asexual budding. The medusa offspring are sexually distinct, being either male or female. The eggs or sperm are shed into the water, where fertilization takes place. The resulting larvae then settle on an appropriate substrate and develop into an attached polyp generation, bringing the life cycle full circle.

There are over nine thousand species of cnidarians, which are classed into three major groupings. In the primitive fernlike hydroids (class Hydrozoa), the polyp is the dominant generation. In the true jellyfishes (class Scyphozoa), the medusa stage is dominant, with a much reduced polyp generation. In the more advanced sea anemone (class Anthozoa), there is only the polyp form.

Whatever the form, attached or free floating, the cnidarians are anatomically simple. There is a single opening for taking in food and rejecting indigestible matter, and there is an internal cavity. The opening is known as

Water jellyfish (*Aequorea victoria*) is a hydroid medusa, usually only 7 centimetres (2¾ inches) in diameter, but it can be larger. The species is easily identified by the sixty-five fine lines radiating from the centre to the edge of the bell.

the mouth, the internal cavity as the stomach. Water circulates freely inside the stomach cavity. Long cilia beat downward, drawing a current into the cavity and providing the anemone with a steady supply of clean, oxygenated water. Other cilia lower down beat in the opposite direction, creating an outward flow to take away metabolic wastes. When the cnidarian is feeding, the cilia beat in the opposite direction to draw food down. This is the cnidarian's version of swallowing.

There is no circulatory system, no excretory system, no respiratory system. The cnidarian body has only two cell layers, forming an inside sheath—the stomach lining—and an outside sheath—the skin. Since each cell is in direct contact with the water, it is able to obtain oxygen and diffuse cellular wastes without complex structures such as kidneys, glands and blood stream.

Between the two cell layers is a cellular secretion known as mesoglea. This is the "jelly" of the jellyfish. The mesoglea is much reduced in the polyp forms, such as the sea anemone.

Covering the cnidarian body is a nerve network providing the animal with a simple nervous system. As a part of this nervous system there are specialized sensory cells in the medusa form. These are the photoreceptors and the statocysts, both located at intervals on the outside margin of the jellyfish body, or bell. The photoreceptors or "eye spots" respond to light, causing the jellyfish to be attracted to or repelled by it, according to the habits of the particular species. The statocysts balance the jellyfish, causing a righting reflex should the jellyfish become disoriented. Because the polyp form is unable to move, it has no balancing or light organs.

The floating medusa moves by a rhythmic pulsing of the bell. By contracting the bell around its margin, the jellyfish forces out the water contained within, providing a weak jet propulsion. From 15 to 150 pulsations per minute occur, depending on the species and the water temperature. Movement is weak and random at best, as the jellyfish has no way of knowing where it has been or where it is going.

Soft, slow-moving, devoid of sense organs to warn of approaching danger, the cnidarians, particularly the jellyfish, present easy prey for hungry predators. Yet these animals have survived 500 to 600 million years in the world's oceans, during which time potential predators have greatly increased in numbers and sophistication. How have the cnidarians survived?

The cnidarians share a not-so-secret weapon—small stinging cells known as nematocysts. These are generously distributed over the body, particularly the tentacles, as oval-shaped capsules containing tiny coiled threads that discharge like small harpoons when stimulated by physical or chemical contact. The stinging cell, which is barbed and poisonous, is just one of eighteen kinds of nematocysts. There are also entangling nematocysts and sticking ones, all combining forces to ensure the capture of any suitable prey (small fish, invertebrates and so on) coming in contact with the tentacles. Animals too large to be paralyzed or killed by the injected toxin would most certainly be stung and thereafter avoid further contact. In this way the stinging cells serve both an offensive and a defensive function. Nematocysts are disposable; once they are used or discharged, they are shed and new ones quickly grow to replace them.

The giant sunfish *(Mola mola)* seems to be insensitive to the nettle cells and is said to feed almost entirely on floating cnidarians. Some sea slugs can ingest the cells without causing them to discharge. The cells migrate to the surface of the sea slug and resume their protective function in favour of the sea slug.

SEA FIRS AND SEA PLUMES (HYDROZOA)

The hydroids are primitive cnidarians. They are often overlooked because of their small size, or they are mistaken for plant growth—hence their popular name of sea fir or sea plume.

One may wonder how a cnidarian animal that is supposed to be cup-shaped could possibly look like a fir branch. The branching hydroid, usually only a few centimetres in height, is actually a whole colony of tiny polyps all joined together, sharing a common digestive canal and stiffened by calcium or some other firm material in the mesoglea. Each colony forms a fernlike structure, the individual members having budded asexually from a single polyp. In the hydroids the polyp colony is the dominant generation. A sexual generation of hydroid jellyfish is budded off from the colonial hydroid

Attached hydroid colonies. Many individual attached polyps are joined to form a hydroid colony. These colonial hydroids occur in a great variety of intriguing and confusing forms.

polyp, but the jellyfish are generally very small and short-lived. The medusa offspring give rise to another hydroid colony.

The hydroid jellyfish are distinguished from the true jellyfish by the presence of a velum, a shelf extending inward from the margin of the bell, and a simple four-chambered stomach. Since the medusae are sexual animals, they contain gonads (sex glands), which are often clearly seen through the transparent body as four distinct clusters, or stripes, just above the four chambers of the stomach.

The water jellyfish *(Aequorea victoria)* is a very common hydroid medusa of the Pacific coast during the summer months. Colourless and almost transparent, this 10- to 15-centimetre (4- to 6-inch) medusa is recognized by the sixty to one hundred threadlike canals radiating from the centre of the bell. Another common hydroid jellyfish is the cross jelly *(Halistaura cellularia)*. The gonads form a distinct cross on the bell, making the cross jelly look as if it is divided into four equal parts. In this species the bell is shallow and is 7.5 to 10 centimetres (3 to 4 inches) across. It is seen from May through to the early fall. Eelgrass beds, where the water is quiet and clear, are the home of the diminutive 12-millimetre (½-inch) orange-striped jellyfish *(Gonionemus vertens)*; the gonads give this species its bright orange color. An occasional visitor that blows in from warmer climes is the blue sail or velella jellyfish *(Velella velella)*. This species has an erect, triangular-shaped sail across the top of the bell.

THE TRUE JELLYFISH (SCYPHOZOA)

In the class of cnidarians known as the Scyphozoa, the jellyfish generation is by far the dominant one. There are two hundred species throughout the world, and they are distinguished from the hydroid jellyfishes by their large (sometimes enormous) size, their lack of a vellum and the often-elaborate development of oral lobes.

In all medusa or jellyfish the mouth extends from beneath the bell by a short, thick stalk, much like an umbrella handle. Surrounding the mouth are liplike flaps known as oral lobes. In the true jellyfishes these lobes are often greatly extended to form trailing streamers or elaborate ruffles. Some even look like pom-poms hanging on ribbons. These structures are in addition to the tentacles found on the jellyfishes' scalloped perimeter.

The simple four-compartment stomach of the hydroid jellyfish is expanded in the true jellyfish into a complex network of digestive channels. This is most probably an adaptation to the greater size of these animals and the need for a more efficient way to supply more nutrients to a larger body mass. Like the hydroid medusa, the scyphozoan jellyfish are sexually distinct, producing sperm or eggs that are shed directly into the water. In some true jellyfishes the eggs are retained and brooded in the marginal folds for a period of time before the young strike out on their own. If a polyp stage occurs (some species skip the polyp stage), the larva settles and develops into a polyp in some area sheltered from strong current and wave action. In the true jellyfishes the polyp is very small, about 12 millimetres (½ inch) in height, and is never colonial. The polyp elongates into a vase shape and begins to divide crosswise, giving the appearance of a stack of saucers. Each saucer separates as a small jellyfish, which swims away and grows to be a big jellyfish.

Just how long this takes, or even how long a jellyfish lives, is not known. It is thought that the complete life cycle, polyp and medusa generations combined, occurs annually, giving each generation a life span of a few months, the dominant generation living the longest. Whatever the case, along the Pacific coast, jellyfish are seldom seen other than in the warmer spring, summer and fall months. One of the most conspicuous of these is the orange-tinted sea blubber (*Cyanea capillata*), a huge (up to 60 centimetres [2 feet] in diameter) jelly mass trailing tentacles of 180 to 240 centimetres (6 to 8 feet) in length. A North Atlantic relative of the sea blubber is *Cyanea artica*, measuring up to 240 centimetres (8 feet) across, with tentacles extending 61 metres (200 feet). The sea blubber is capable of causing a nettlelike sting, even when dead and washed up on shore. Contact with the skin, particularly in children, should be avoided. The moon jelly *(Aurelia aurita)* is a smaller jellyfish, usually 10 to 15 centimetres (4 to 6 inches) in diameter, common to bays and estuaries. It is colourless except for the four horseshoe-shaped gonads coloured violet, pink or yellow. The oral lobes of this

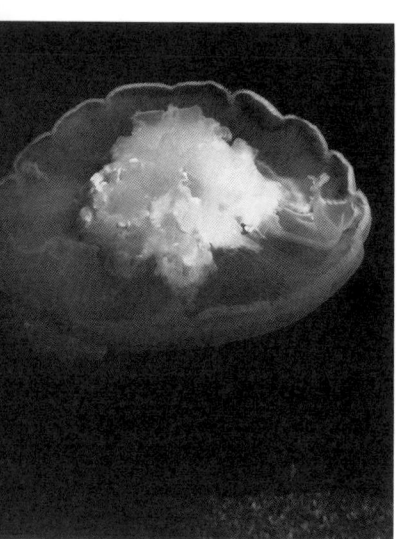

The moon jellyfish (*Aurelia aurita*) can be very large, up to 40 centimetres (15¾ inches) in diameter, but it is generally smaller. Fringed oral lobes on the animal's underside are easily seen here. Not shown are the four distinctive horseshoe-shaped gonads, which can be seen when you are looking down on the jellyfish. The moon jellyfish feeds on plankton such as copepods, which become entrapped in mucus on the bell and oral lobes and then move to the animal's mouth along tracts of cilia.

species are large, whereas the tentacles are short and delicate.

A good way to watch jellyfish is from a dock or wharf. Stretch out, belly down, head and shoulders hanging over the water, and watch the pulsing movements of the animal. The jellyfish isn't in a hurry, and presumably you aren't either.

Another summertime method is to take a fine mesh dip net, the kind you buy in a pet shop for aquarium fish, and wade out into an eelgrass bed. Dip the net in a kind of shovelling motion, spade in and pull up. In so doing, you will scoop from the bottom to the top of the water and capture anything in the net's path, including small jellyfish.

Yes, you can touch a jellyfish on the West Coast. The ones that inflict terrible stings (the sea wasps, cube jellies and Portuguese man-of-war) await you on your next tropical vacation. Just as a precaution, though, I would not recommend that very young children with delicate skin play with jellyfish, dead or alive.

SEA ANEMONES, CORALS AND SEA PENS (ANTHOZOA)

The Anthozoa (the sea anemones, corals and sea pens) have completely forsaken the medusa stage of the hydroids and jellyfishes, retaining only the attached polyp form. They are by far the most conspicuous and successful of the cnidarians, having over six thousand species, two thousand of which are anemones.

The sea anemones share with some jellyfishes a peaceful beauty and delicacy, an esthetic loveliness of radial symmetry, waving tentacles and colorful hues. It is difficult to believe that these simple, almost stationary animals are carnivorous and, in some cases, capable of doing battle with their own kind.

The sea anemones are an improvement on the basic polyp form. The gut cavity is divided into vertical sections by membranes known as septa, which extend from the body wall inward towards the centre of the gut. In most species there is a distinct bottom, or pedal disk, which attaches the animal firmly to its chosen substrate. The column, or stalk, of the thick body is crowned with tentacles surrounding an oral disk. The disk has an elastic mouth opening in its centre, which acts as a kind of permanent lid, closing the end of the tubular body.

Tentacles can number from a few dozen to thousands. Anemones feeding on larger prey such as unwary crabs, invertebrates or small fish, tend to have fewer, stouter and stronger tentacles than those feeding on tiny planktonic organisms. The latter type of anemones generally have great fluffy masses of very fine, thread-like tentacles, as seen in plumose anemones *(Metridium senile)*.

By and large, anemones are not considered mobile creatures, though an anemone can move very, very slowly by sliding on its pedal disk, or foot, if environmental conditions indicate a change. This movement may be in response to a more abundant food supply or a need for greater shelter or less direct sunlight to avoid drying out if exposed at low tide. *Condylactis* is the sprinter of the anemone set, moving at 25 centimetres (10 inches) per hour, followed by *Sagartia* at 2.5 centimetres (1 inch) per hour and *Metridium* at 12 millimetres (½ inch) per hour.

Some species of anemones are chauffeured around the sea floor attached to the shells of crabs and mollusks, and one anemone is able to "swim." *Stomphia coccinea* is one of the few anemones that will actually flee

Cross-section of an anemone.

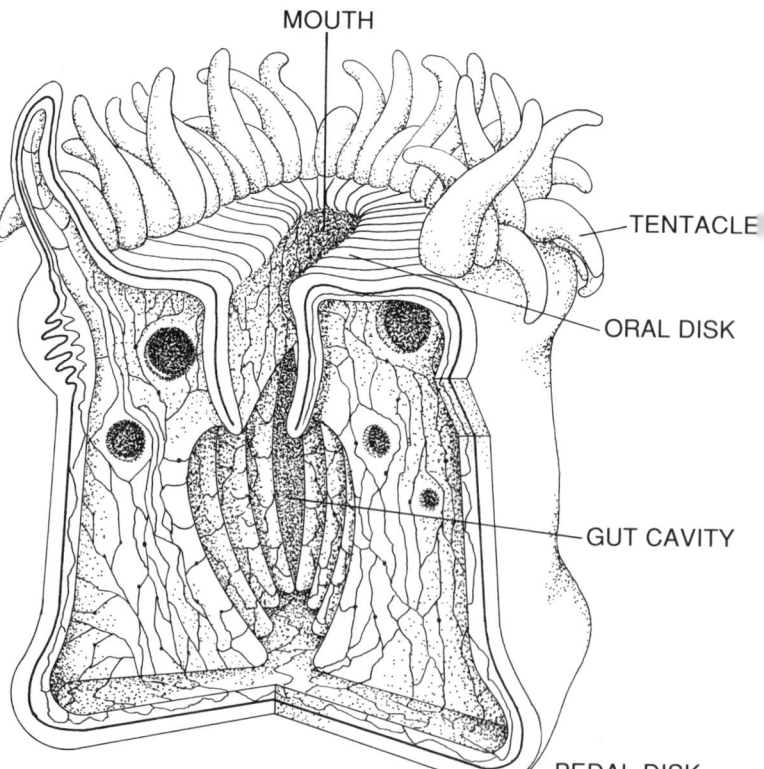

MOUTH

TENTACLE

ORAL DISK

GUT CAVITY

PEDAL DISK

its enemy, the leather star *(Dermasterias imbricata)*. Sensing the presence of a leather star by means of sensory cells on the tentacles and body, the *Stomphia* will release its foothold and by alternate contraction and relaxation of the body column muscles, wriggle away to safety.

Without specialized and efficient locomotory adaptations, the anemones are unable to outrun the majority of potential predators. They have, instead, developed a twofold defence system: stinging cells and size reduction.

Like all other cnidarians, the sea anemone has nematocysts, or stinging cells, causing most fish to diligently avoid contact. For predators undaunted by nematocysts, the anemone attempts to reduce temptation by withdrawing its upper body inward, folding its tentacles and oral disk into its body cavity, and thereby reducing itself to an igloo-shaped lump. Some species attach bits of sand and shell to their body columns, further camouflaging their presence when tucked in. Even so, anemones fall prey to a variety of predators: certain mollusks, starfish, flatfishes and some other fishes.

Anemones may also fall prey to other anemones, not as food items, but as a result of a territorial action. Some discourage newcomers that approach too closely by inflating small, saclike protuberances at the base of its tentacles and leaning over to touch the outsider. Contact with these poison sacs will destroy tissue on contact and even kill the interloper if it does not retreat beyond the tentacles' reach.

Although nematocysts are effective as a defence mechanism, they are primarily a predatory adaptation used in capturing food. The effect of discharged nematocysts can be felt as a kind of stickiness if you put your fingers into the tentacles of a green surf anemone *(Anthopleura xanthogrammica)*. Once prey has been entrapped or stunned by the stinging cells, food items are transferred by the tentacles towards the anemone's mouth and into the gut cavity for digestion. Indigestible material is ejected through the same opening.

One of the most intriguing aspects of anemones is their reproduction methods. The anemone has not restricted itself to a process of simple pairing but has a number of reproductive options. Anemones may be hermaphroditic, having both male and female gonads, or they may be dioecious, having only male or female

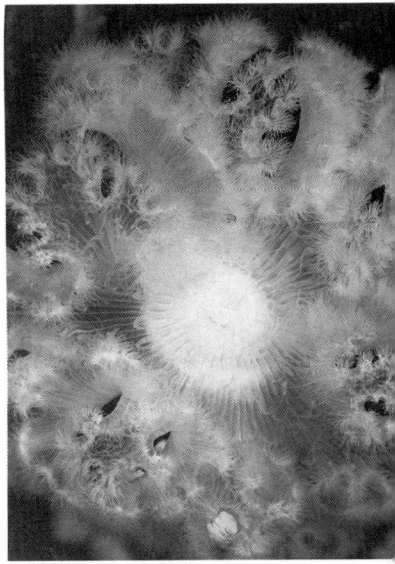

The plumose anemone (*Metridium senile*) occurs in a variety of colours: white, orange and brown. Some may grow very large, and when the stalk is extended they may reach 90 centimetres (3 feet) tall. Tentacles are numerous, very fine and worn in a fluted margin. Plumose anemones are particle feeders, capturing tiny waterborne organisms in a mist of tentacles. Plumose anemones are frequently encountered on floats, in the low intertidal and in deeper water.

Cnidarians and Comb Jellies   33

gonads. The eggs and/or sperm are shed directly into the water through the mouth, tentacle ends, or special pores on the body. Some species even brood their fertilized eggs.

Plumose anemones (*Metridium senile*) are also capable of asexual reproduction. With one such method, the anemone spreads its pedal disk, then draws in the central portion, leaving a ring of tissue shreds that were the perimeter of the anemone's base. Each bit so left develops, or regenerates, into a new anemone. Another method of asexual reproduction is accomplished by a vertical splitting. A furrow develops on either side of the anemone, running from base to summit. Over a period of hours or days, the furrow deepens until the anemone is split into two complete animals.

This kind of reproduction raises some fascinating problems about life and death, at least in anemones. Are the resulting offspring of a split anemone really offspring, or are they duplications of parents? Can this really be considered a new generation? From what point would one begin to calculate the life span of the individual?

Sea anemones are incredibly long-lived if one can generalize from the few available documented cases. One of the most famous of these is a batch of anemones collected sometime before 1862 and tended by a woman who changed their water and fed them on fresh liver. Eventually these specimens were given to the Department of Zoology at the University of Edinburgh, where they continued to thrive until they were all found dead in the early 1940s. The cause of death was thought to be human error, but whatever the cause, the anemones were at least eighty years old. Some biologists believe it is possible, and even probable, that anemones can live to at least one hundred or even a few hundred years.

Anemones are also among some of the deepest-living animals; specimens have been dredged from depths of 9000 metres (29,528 feet) in the Philippine Trench (1952). They also occur high on the intertidal area and withstand the rigours of exposure experienced by all intertidal animals and plants. They range from a few millimetres to over a metre across in the enormous tropical carpet anemones. Two species of surf anemones are abundant intertidally on the rocky shores of the Pacific Northwest. The smaller of the two species (*Anthopleura*

Brooding anemone (*Epiactus prolifera*). After remaining under the protective tentacles of its parent for about three months, a young anemone moves off the parent's stalk but remains nearby. Consequently, you will often find a number of this species together. Look for this anemone under rocks, on algae and on blades of eelgrass. It is small, about 5 centimetres (2 inches) across, colour is variable, and it has fine stripes on the stalk. Leather sea stars and shag rug nudibranchs are predators on the brooding anemone.

elegantissima) occurs in large beds, often in tide pools, whereas the larger species *(Anthopleura xanthogrammica),* which is up to 20 centimetres (8 inches) across, occurs singly and more towards the subtidal region. The deep green colour of the latter is due to the presence of a commensal alga growing in the cells of the anemone. If deprived of sunlight for a period of time, the infecting plant dies, rendering the anemone white.

Another very common intertidal anemone is the 2.5-centimetre (1-inch)-high *Epiactis prolifera.* Although named "small green anemone," it can be green, red or brown, and it can be covered with camouflaging debris.

Large-plumed *Metridium* anemones, up to 30 centimetres (12 inches) in height, have the dense, thread-like tentacles of plankton feeders. Specimens vary from white to orange to dark brown and are found in very deep waters as well as near the low tide mark. Large *Tealia* anemones, which have fewer, thicker and stouter tentacles than the fluffy plumose variety, are red to pink, with beautiful banding patterns on the tentacles.

Anemones are nearly always photographed or illustrated as they appear under water, with their tentacles waving gracefully in the water. There is a reason they are shown in this way. Under water they are beautiful; exposed at low tide they are flaccid, igloo-shaped blobs. If they hung their tentacles out in air, they would dry out. Thus, the anemone folds its tentacles into its mouth and relaxes until the tide comes in. This means that when you are looking for an anemone at low tide, you will look for something that has a low profile and is shaped like an igloo unless the anemone is hanging upside down from the undercut of a ledge, looking rather obscene.

If you think you have found an anemone, give it a gentle poke with your finger and it should respond by contracting under your touch. Don't worry; it won't hurt you. Even more fun is finding an open anemone in a tide pool. Gently touch the tentacles and you will feel the sticky cells that help the anemone entrap its food.

Surf anemones (*Anthopleura elegantissima*) are common intertidally on rocky shores, often in large groups. Individuals grow to 4 or 5 centimetres (1½ to 2 inches) in height and are often seen with bits of sand and shell attached to the stalk. The latter helps to keep the animal from drying out when it is exposed at low tide. During low tide look for sand-covered doughnuts of jelly in channels. These are probably surf anemones.

SEA PENS

The sea pens, named for their resemblance to the plumed quill pens of old, look and behave as a single organism, yet they are not. The sea pen is a colonial animal and is placed in the same group as anemones and corals. Each "leaf" of the plumed portion of the sea pen has many tentacled feeding polyps on its leading edge. The structure and function of the individual polyps are closely allied to the polyp form of the anemone and coral. The stalklike body, plumed on its top half, is supported by an internal stiffening rod, allowing the sea pen to stand upright on the sandy bottoms where it lives. The foot end digs into the sand by peristaltic (wormlike) movements, the result of alternate expansion, with water, and contraction. Once buried, the foot end extends itself with water, creating a firm anchor for the upright sea pen. Sea pens have a reasonable degree of mobility and are able to withdraw their bulbous end, moving on to a more favourable environment if conditions warrant a change.

Ptilosarcus guerneyi is approximately 50 centimetres (20 inches) when fully extended and a mere fraction of that when contracted. It ranges in colour from pale to deep orange and is found subtidally over sand and mud bottoms. A heavy slime secretion bears luminescent granules, causing the sea pen to glow in the dark. The sea pen is preyed upon by the striped nudibranch (*Armina californica*) and the pink nudibranch (*Tritonia festiva*).

CORALS

Most people are as surprised to find coral along the West Coast as they are to find eleven species of shark in this area. Both sharks and corals are usually associated with warm, tropical seas.

It is true that none of the reef-building corals—those responsible for tropical coral reefs—occur in temperate waters. A few coral species do occur here, however. Two of these are solitary, stony corals looking like tiny anemones that have secreted stony cups to sit in. The cup is actually a skeleton. Both species are small, 6 to 12 millimetres (¼ to ½ inch) in size, and range in colour from red to orange or yellow to grey. The orange solitary coral (*Balanophyllia elegans*) may be found in pro-

Sea pen (*Ptilosarcus guerneyi*).

Solitary coral (*Balanophyllia elegans*). This is the only true coral to occur intertidally on the West Coast. The animal can withdraw completely into its 1-centimetre (½-inch)-wide stony cup when exposed at low tide. A telltale sign to look for is the bright orange to yellow colour in the cup, even when the tentacles can't be seen.

tected areas or under rocks from the low tide level to 46 metres (151 feet).

A beautiful deep-water "soft" coral, known as a gorgonian or pink candelabrum coral *(Paragorgia arborea)*, forms a colony branching upward from a central stalk—hence the name candelabrum. In addition to a stony skeleton, the gorgonian coral secretes a woody material, giving rigidity to the 45-centimetre (18-inch)-high colony. The pink candelabrum coral is found in waters that are 60 metres (197 feet) deep or more.

SEA GOOSEBERRIES OR COMB JELLIES (CTENOPHORA)

Perhaps the beachwalker may come upon little oval balls of jelly stranded on the sand that usually look very much like small jellyfish. These are usually ctenophores, or comb jellies, commonly called cat's eyes or sea gooseberries. The comb jellies, though similar in appearance to jellyfish, belong to an entirely different group and are absolutely unrelated to any cnidarian. They do not pulsate as medusa-type jellyfish do but are moved through the water by rows of beating hairs, like the teeth of a comb—hence the name comb jellies. Some comb jellies have tentacles. These can be extended twenty times the length of the body and are used to capture eggs, larval forms and small fish. There is no alternating of generations.

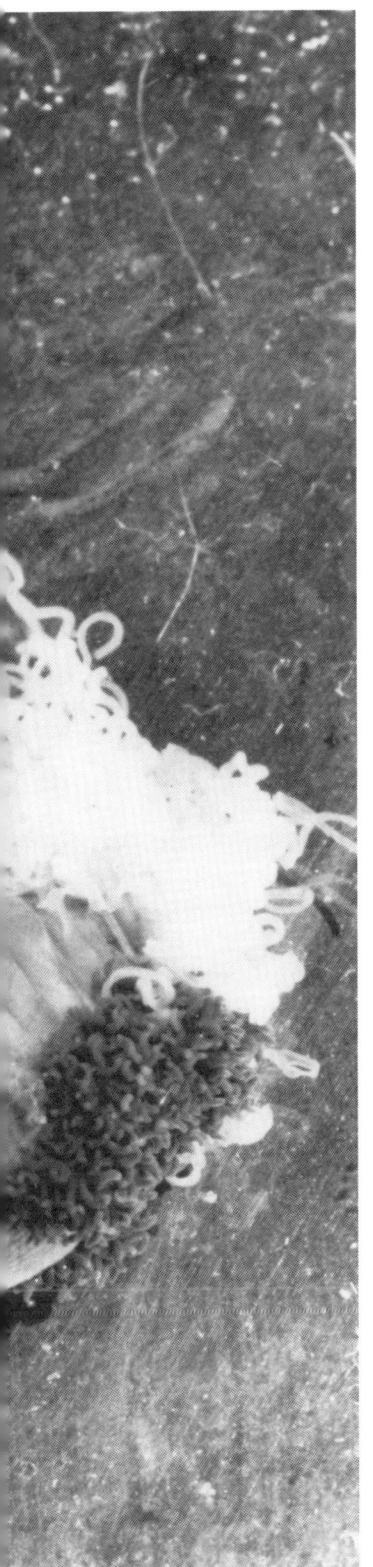

CHAPTER 4

Seashore Worms

One of the great surprises of poking around in mussel beds when the tide is out is finding a long, skinny bright red worm. It doesn't do anything exciting; it's the colour that is such a treat. Do look for it. Or turn over some good-sized rocks and try to find a coiled white, limey tube. You will not see the worm that lives there, since it will have sucked itself as deeply as it can into the tube, but you will have the satisfaction of knowing that you have at least found a serpulid worm, even if you haven't seen it.

If you want to find out about shipworms, you have come to the wrong place. Shipworms, also known as teredos, are not worms; they are clams! You will find the fascinating story of teredos in the chapter on mollusks.

FLATWORMS (PLALYHELMINTHES)

All flatworms are not flat; nevertheless, those most easily recognized in nooks and crannies at low tide on a rocky beach will inevitably be flat. Intertidal flatworms are commonly called leafy flatworms because they resemble small leaves in shape and thinness.

Although in this book flatworms have been placed together with other worm forms for the sake of continuity, in evolutionary development the flatworms belong right after the comb jellies or ctenophores. In their physical development, flatworms are not as sophisticated as the segmented worms. For example, flatworms have no specialized respiratory organs or tissues, such as gills, for the absorption of oxygen. There is no circulatory system, such as heart or blood vessels, to distribute nutrients and supplies to the cells. There is not even an anus to dispose of indigestibles, which must leave the way they came in, through the mouth.

A flatworm typically has a mouth, a sac for a stomach and, surprisingly, complex reproductive equipment of both sexes, a development that appears out of proportion to that of the rest of the organism. In short, the flatworm is bent on eating and reproducing and little else. It is not surprising, then, that two of the three classes of flatworms have evolved a totally parasitic way of life. Examples of such flatworms are the dreaded tapeworms and various internal flukes that infest both humans and beasts. However, the parasitic flatworms are not to be confused with the free-living polyclad flatworms of the intertidal region.

Flatworms are considered to be the first group of many-celled animals to show bilateral symmetry, or indications of a distinct right and left side in relation to a distinct front and back; that is, one side is a mirror image of the other. This bilateral symmetry is considered an improvement over the radial symmetry of jellyfishes, anemones and hydroids, suggesting that the flatworms have advanced over the anemones et al. Polyclad flatworms, so named for their many-branched stomachs (*poly* = many, *klados* = branched), move over rocks or mud by gliding on a track of secreted slime, using backward sweeps of the cilia that cover the worm's lower surface. When potential prey is recognized, a portion of the pharynx is everted from the flatworm's mouth,

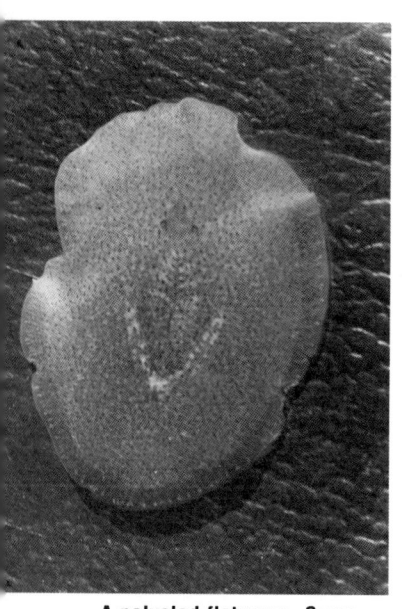

A polyclad flatworm. Some flatworms, such as flukes and tapeworms, are parasites, and some species are free living and nonparasitic. On the beach they are recognized by their thin, leaflike bodies and are best found on the undersides of rocks or boulders turned over during low tide. They can also be found on seaweeds and in tide pools but are more difficult to see there.

which is located in the middle of its underside. Descending on its meal like a limp umbrella, the flatworm engulfs small creatures such as other flatworms, tunicates and shrimplike animals.

The polyclads are strictly carnivorous, eating only animal material. One small polyclad has been observed feeding on barnacles, which it harasses by crawling up the side of the barnacle's shell and apparently secreting some irritant or poison into the barnacle's interior. The barnacle responds by waving its cirri, or tentacles, as if trying to swat an elusive fly. Eventually, the barnacle is overcome and the flatworm dives into the barnacle shell to consume the now moribund animal.

RIBBON WORMS (NEMERTEANS)

Ribbon worms are generally not found crawling around in plain view, though they are certainly common in rocky areas or wherever large numbers of attached animals and plants occur to provide the worms with both cover and food.

Ribbon worms are typically long and thin, and unlike the annelid worms, which show strong segmentation, the ribbon worms are smooth, soft, slender and very elastic. They are fragile creatures that tend to break apart easily if handled. In fact, the ribbon worms give the impression of being not quite "done." They appear to lack all the stuffing they should have, causing them to be slightly flattened or flaccid looking. Their skin is so soft that it seems to have not quite finished toughening and appears to need some kind of protection. In fact, some ribbon worms do secrete a papery tube about themselves.

A ribbon worm (*Amphiporus bimaculatus*). The head bears a pair of dark wedge-shaped marks.

A number of ribbon worms are common to the West Coast. These range in size from a few centimetres to many metres long and in colour from deep blood-red to brown. Or they may have contrasting bellies and backs or bandings of various colours.

Not only do ribbon worms lack the segmentation and parapodia of the annelid worms, they also lack the biting jaws. Instead, the ribbon worm captures its prey of polycheate worms and small shellfish by using what is called an eversible proboscis, which is used in much the same way as a frog's tongue. When the worm shoots out its proboscis, prey is captured by the gluey or barbed end, poisoned to calm its frantic efforts to escape, and then eaten.

ANNELID WORMS

You may think you are at just another rocky beach with bits of seaweed here and there, a few barnacles and, near the tide line, some crab casts and broken shells. But turn over a boulder at low tide, and you will see an entire microcosm of marine life. After the crabs and eel-like fishes have beat a frantic escape, look carefully at the underside of the turned boulder. You might see some small, clear, jelly blobs that are perhaps transparent sea squirts. Some milk-white, sluglike creatures are most likely sea cucumbers, but who or what is responsible for the hard-coiled white tubes firmly glued to the rock's underside? The tube is the home of a small serpulid worm, only one of at least six thousand living species of marine annelid-type worms.

The name Annelida refers to a diversified group of worms. It comes from the word *anellus*, meaning "little ring," and refers to the repeated rings around the worm's body—as in the familiar earthworm. Each ring delineates a segment, the building block of the annelid body plan. Aside from the head and tail segments, all of the body segments, at least in the more primitive annelids, are essentially the same, inside and out. For example, each segment has its own pair of kidneys and its own nerve-exchange site. Like spools, the segments sit on one another and are open in the centre to allow for the continuous passage of the digestive tract. Unlike the earthworm, most of the marine annelid worms also have side flaps with bristles on each segment. These parapodia, as they are known, generally serve two functions: movement and respiration. They are, therefore, feet and gills in one.

The fact that the parapodia of some polycheate (meaning "many bristles") worms hang limp and ragged when removed from the sea has earned this group of worms the name rag worm. They are more correctly termed nereid worms.

A great variety of free-living marine worms of the kind described above can be seen at the seashore—in mussel beds, in eelgrass or in sand turned up while you are digging for clams. The worms can be quite large, up to 60 centimetres (2 feet) long, have beautiful iridescent

A scale worm (*Halosyda brevisetosa*). A good place to look for a scale worm is between the shell and foot of a limpet. The scale worm gains protection and a free ride from this arrangement, but it is not a parasite on the limpet. Other scale worms live as commensals in the tubes of tube worms. Free-living scale worms of this kind are about 6 centimetres (2¼ inches), whereas those living with the tube worm are twice as long.

colouring and a nasty bite. The biting jaws are hidden back in the throat when not in use but can be pushed out and against some likely prey when needed. Nereid worms are active, predatory creatures that feed on other worms, small crabs and shrimps, a variety of larval forms and seaweeds. Nereids make wonderful fish bait and can easily be found in clumps of mussel. The head should be removed before the worm is placed on the hook for obvious reasons.

Nereid worms, for all their fierce and predacious nature, conceal themselves under rocks or in sand for the same reason that makes them excellent fish bait. They are attractive food for a great variety of fish as well as some marine mammals and birds.

Among the huge array of polycheate worms, a number of species have forsaken the active life of a free-living worm, and its incumbent dangers of predation, for the restriction and security of a fixed life within a protected house of the worm's own making. Hence the worm of the white, coiled house under the rock.

The homes of such worms are known as tubes, and the worm within is called a tube worm. There are two common kinds. Those that secrete a limey, hard tube are known as serpulids and are generally small. Those that secrete a leathery or parchment tube are named sabellids and tend to be larger. One sabellid (*Eudistylia vancouveri*) builds a tube up to 50 centimetres (20 inches) long with a diameter of 12 millimetres (½ inch). The worm itself, however, at 15 centimetres (6 inches), is only a fraction of the tube's length.

In assuming a tube-dwelling existence, the tube worms have been required to alter their feeding habits drastically. Unless the worm can feed within the security of its home, the benefits of building a house would be greatly diminished, since the worm would have to forage away from the tube and therefore present itself as easy prey. In response to the food problem, some tube worms have developed beautiful plumed "feathers" at their head ends, which they can expose to the water from the tube's open end. Fine hairs on the plumes, covered with mucus, trap floating particles. The mucus and entrapped food are then swept towards the mouth by beating of the hairs and swallowed. Because the plumes of the leathery-tubed worms (sabellids) are much more branched than those of the serpulid worms,

Nereid worms are nearly everywhere at the seashore. They range in size from small, inconspicuous species to the very large ones, in excess of 80 centimetres (31½ inches). As annelid, polycheate worms, they are generally easy to distinguish by their segmented bodies and paired bristles or flaps on each segment, running in a continuous line down the sides of the worm.

Parchment tube worm (*Eudistylia vancouveri*) as seen at low tide. These worms characteristically occur in large clumps of many individuals. When the plumes are exposed, the mass looks like a great bouquet of chrysanthemums. Individual plumes can be 5 centimetres (2 inches) in diameter and are generally coloured a deep maroon with green bands. The tubes housing the individual worms may be over a centimetre (½ inch) in diameter and 25 to 60 centimetres (10 to 24 inches) in length. Rocky, intertidal areas and pilings are the preferred habitat of the species.

Calcareous tube worm (*Serpula vermicularis*). A coiled or rambling tube of white, limy substance found adhering to a substrate indicates the presence of a calcareous tube worm. If the small red cirri are not in evidence, the red operculum or "stopper" at the open end of the worm's tube will tell you that a living worm is within. This species is found intertidally in rocky areas, where it will remain in water even when the tide is out, on the undersides of boulders that retain water in the depressions. Specimens are easily found by looking on the undersides of large boulders turned over at low tide.

the sabellids are commonly known as feather duster worms. The plumes not only capture food but also absorb oxygen, thus serving as gills. Eye spots at the base of the feathery structures are so sensitive that even a shadow passing over the plumes will cause the worm to withdraw them instantly into the tube. Calcareous tube worms, which are serpulids, have even developed a "stopper," which is held against the tube's opening after the worm has withdrawn. Should a fish or some other predator be quicker than the "eye" of the tube worm and chance to bite off part of the feeding structure, the tube worm is quickly able to regenerate its lost parts.

Tube life imposes some additional problems on the tube-dwelling worm. Since the tube is closed at its posterior end, the worm must arrange for the elimination of digestive wastes. In some tube worms, a cilia-lined channel has developed to sweep wastes from the worm's posterior end forward to the tube's open end and out, keeping the tube free of waste material. In the same way, eggs and sperm are released to the open sea to be fertilized by chance. Since tube worms often occur in large aggregations and spawn simultaneously in response to some complex rhythm, much of the chance is eliminated. Even the free-living nereid worms, which are not restricted by life in a tube, do not copulate but come together in groups at spawning time, releasing eggs or sperm through ruptures in the posterior body segments.

With so many worms inhabiting the marine environ-

ment throughout the world, from the intertidal zone to waters many hundreds of metres or more in depth, no clear and absolute definition can be drawn between the free-living nereid worms and their modified cousins, the plume worms. A whole host of forms exhibiting all manner of peculiar modifications occurs in between. For example, the lugworm *(Abarenicola pacifica)* digs a U-shaped burrow but secretes no tube. This worm is responsible for the coiled sand casts found on the sand surface of quiet bays. The casts are the feces that result as the worms eat their way through the substrate. Another strange worm is the terebellid *(Thelepus crispus)*. This animal puts together a soft mud tube, but instead of plumes, it spreads many fine, long tentacles over the sea bottom to collect detritus and particles of food that happen to rest within the tentacles' reach.

PEANUT WORMS (SIPUNCULIDS)

Another worm of rocky shores is the peanut worm (phylum Sipuncula), so named for its resemblance to a shelled peanut when in the contracted state. When extended, the 5-centimetre (2-inch) peanut transforms into a 10-centimetre (4-inch) worm, narrow at the head end but bulbous at the hind end, like a baseball bat.

Peanut worms are found intertidally on rocky shores under rocks or in crevices where sand and mud have become deposited. The drab colour of two common West Coast species *(Phascolosoma agassizzi* and *Themiste pyroides)* and their burrowing habits make these worms something of a challenge to locate. A careful sifting of mud out of some protected pocket will generally yield a smooth, leathery, tough peanut worm.

The peanut worm lacks rings or segments and jaws. Instead, it has fine, mucus-covered tentacles, which are extended out of the mouth when the worm is feeding on suspended particles or detritus in the sand and mud.

As an adaptation to living in a burrow, the peanut worm's anus is located approximately one-third of the animal's length from the mouth. This allows the worm to eliminate wastes without fouling the burrow or having to turn around. Peanut worms are either male or female but do not mate. Eggs and sperm are shed into the open water, where fertilization occurs and results in a small larva. After a period of free living, the larva settles and becomes a regular peanut worm.

Peanut worm (pylum Sipuncula). Although some species of peanut worms can grow to 60 centimetres (24 inches), most are much smaller at 2.5 to 15 centimetres (1 to 6 inches). They are tough and leathery to the touch and can be found wedged in crevices of rocks, among the roots of eelgrass or burrowed in gravel sand or mud in an attempt to avoid being eaten by a hungry fish or a predatory snail.

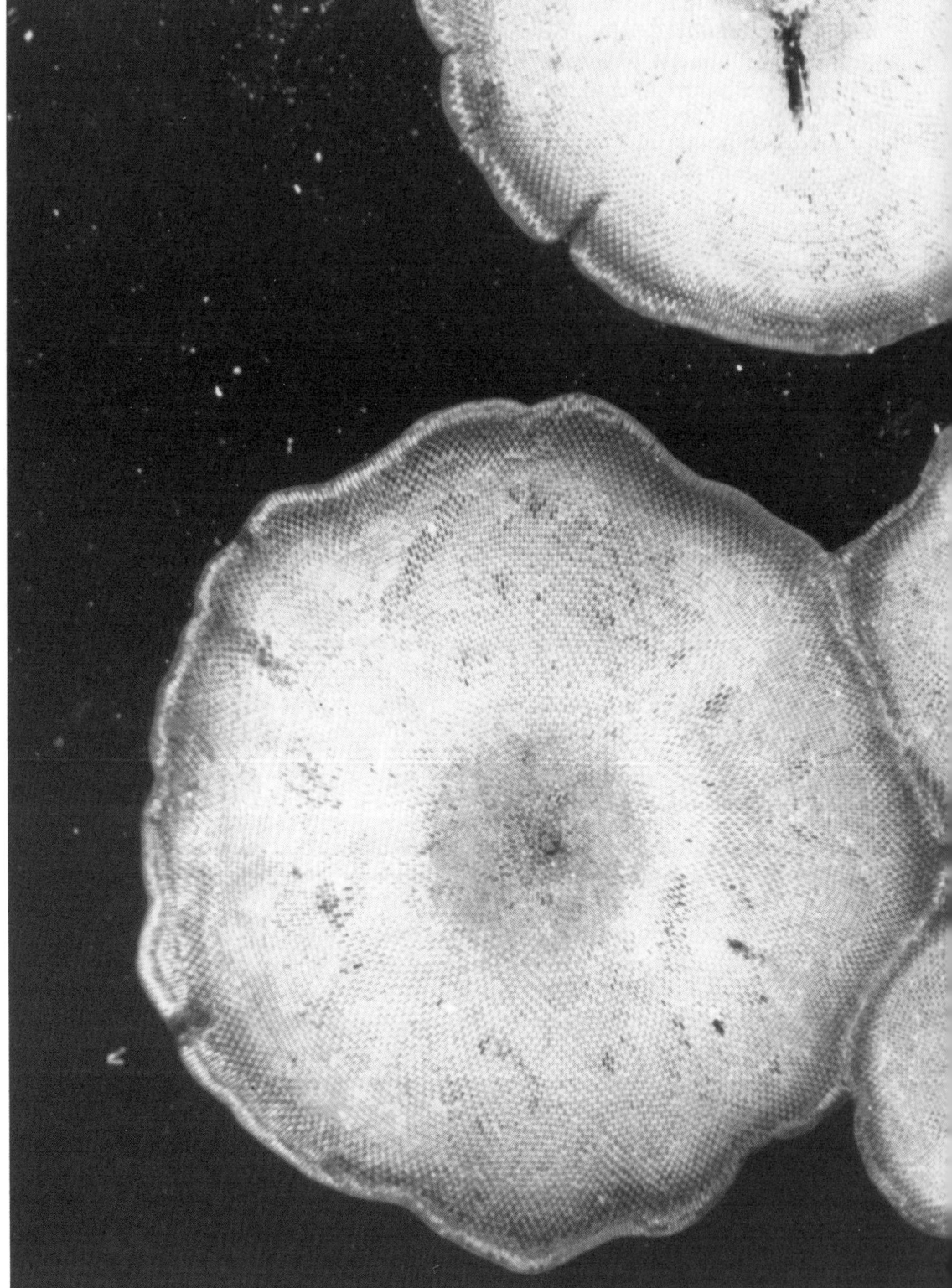

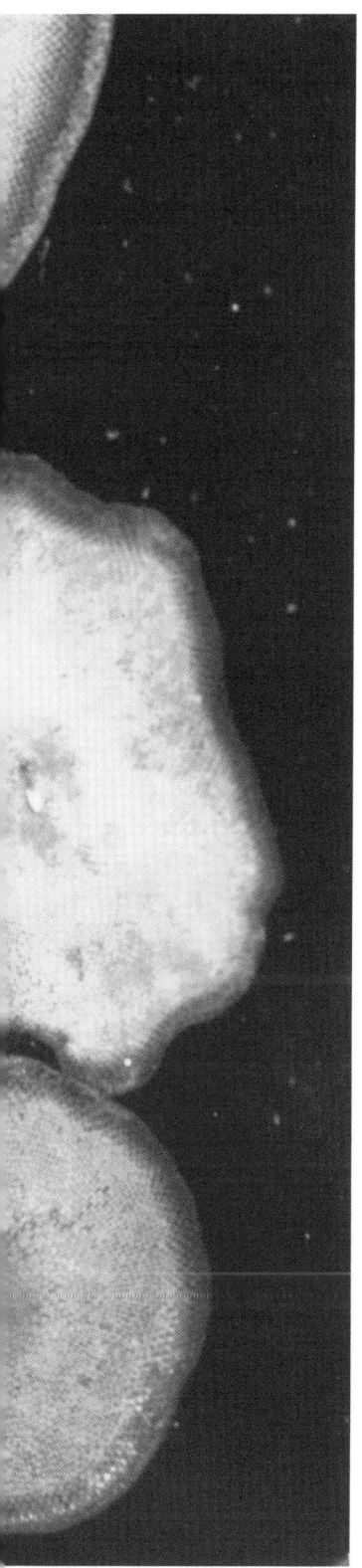

CHAPTER 5

Moss Animals

For some extra fun on your beachwalk, tuck a clear glass bowl into your lunch bag or camera bag to use as a mini-aquarium. (Pyrex or some other type of sturdy glass is best.) When you come across something that looks promising, you can fill your bowl with water and float your treasure in it. You will be amazed at how different things will look in water; they also behave differently. Fuzzy bits may turn out to be animals with tentacles. Worms may relax and show you their mouths and jaws, and small shore crabs will walk around looking for escape. If you pack a magnifying glass with a handle as well, you will be able to see even more details.

Bryozoans, commonly known as moss animals, will be much easier to identify if you float them in your bowl, and interesting details will be easier to see if you use your magnifying glass. For beginners, the fastest way to find some bryozoans is to look for them in the long blades of bull kelp. Large white circular patches that look like a finely crocheted doily are colonial bryozoans. Now watch carefully for the emergence of feeding tentacles, proof that you have found an animal and not a plant.

It seems a paradox to describe bryozoans as abundant and obvious seashore animals when individual bryozoans average less than a millimetre in size. In fact, only a few species occur as solitary individuals; most are colonial forms. That is, many, many individuals live connected together in a large mass, resembling a miniature condominium, each in a separate little shell-like house. As colonies, the bryozoans assume such diverse shapes as to defy definition. Some take the form of low, bushy mats that look like mats of moss—hence the name "moss animals." Or, the colony may resemble a thin, branching, leafless growth a few centimetres in height. Others mimic the shape of some corals, and yet others form circular silvery patches of lacework on the broad blades of brown seaweed. In short, the bryozoans are nearly impossible to describe as a group, even for the expert. One almost firm rule can be stated, however, and that is that bryozoans are generally attached to something: rocks, wood, plants or animals. The ever-present exception is a tiny bryozoan that lives free among sand grains.

Individual bryozoans bear superficial resemblance to coral polyps or miniature anemones. The body is sac-like, with a circle of feeding tentacles around the mouth. Rather than one entrance to the stomach that acts as both a mouth and anus, as in the coral polyp or anemone, the bryozoan has a complete looped digestive tract that opens to a distinct anus located outside the ring of tentacles. A secreted external skeleton-house protects the animal; it can withdraw completely into the house when disturbed. Bryozoans have no circulatory or respiratory system and no nettle cells, as the corals do. In some colonial species, individuals bearing whips or jawlike structures are positioned on the sides of their houses, presumably to discourage other settling organisms from establishing themselves on top of a colony of bryozoans and, in doing so, smothering it.

Brachiopod or lamp shell (*Terebratalia transversa*) superficially resembles bivalved mollusks such as clams and oysters. However, the two are in no way related; in fact, brachiopods are closely related to bryozoans. There are two shells, to be sure. These form the back (dorsal) and belly (ventral) shields of the animal within, as distinct from the two-shelled mollusks, where the shells cover the animal from side to side and are hinged together dorsally. Brachiopods are typically attached to a firm substrate by a flesh stalk emerging through a hole or groove to the back. The stalk, or peduncle, is retractile and able to rotate the shells. There is often no hinge, as in the clams; the shells are held together by muscles.

Suspended food particles are brought to the brachiopod by water currents and are trapped in what looks like a double coil or spring covered by rows of fine hairs, a structure known as the lophophore. In some, fertilization and early development of young takes place in a brood pouch within the female. Free-swimming larvae are soon released and after a brief, mobile existence settle and metamorphose into the adult form.

Illustrated here is only one of a number of West Coast species. However, this is the most commonly encountered and occurs intertidally. The valves may be heavily ribbed or smooth, tending to give the appearance that there is more than one species.

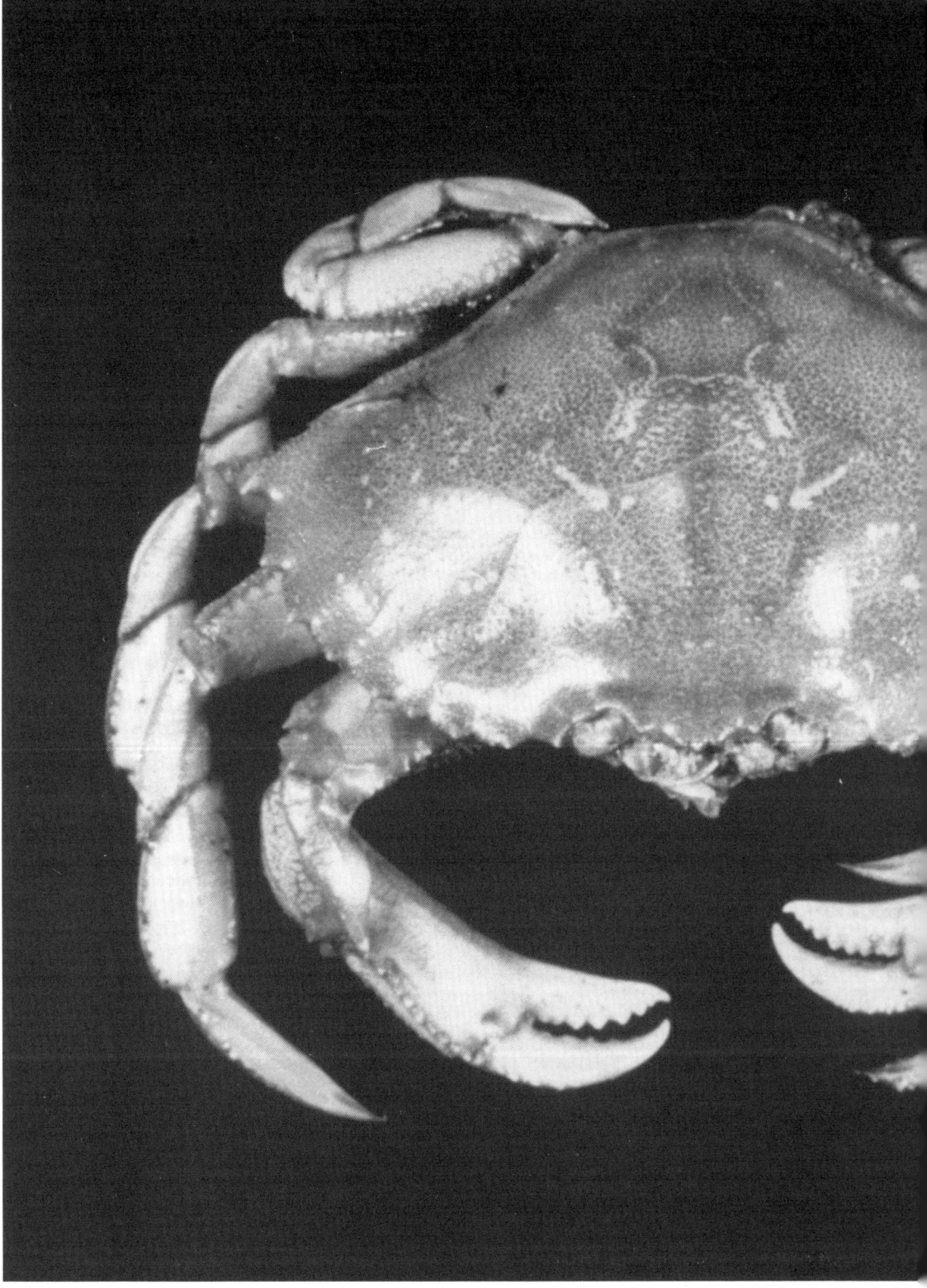

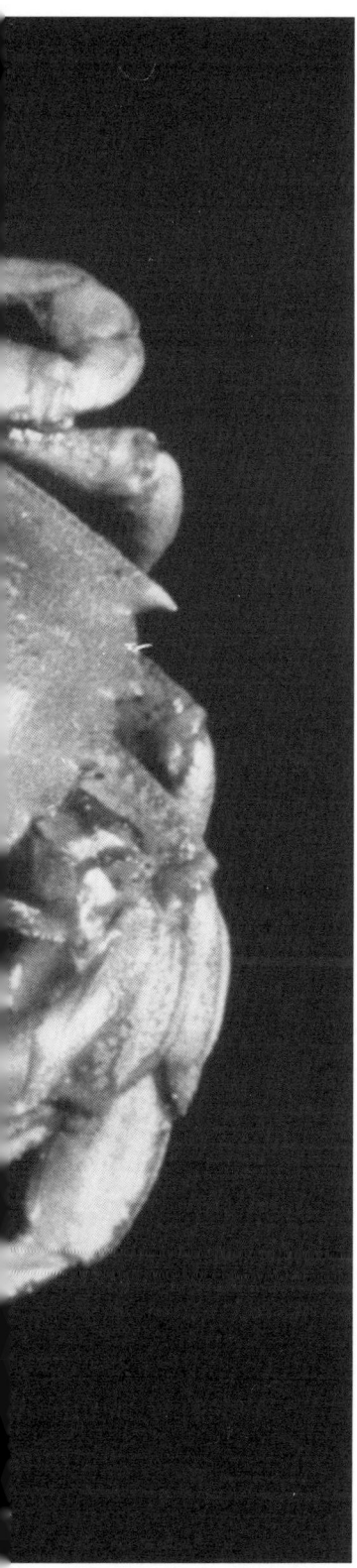

CHAPTER 6

Crustaceans

Crabs need little explanation. Just about everyone knows what they look like and that they have pincers. All crabs are edible.

Sometimes the beach looks as if it is littered with dead crabs. Many of these crabs may actually be molted crab shells. Here is how you can tell the difference between a molted crab and a dead crab. If the top shell lifts open like the lid of a tin can, leaving the bottom of the shell and the legs attached, you have a molt—the cast-off shell of a living crab. This is a good sign because it means the crabs in the area are alive and well and growing. They will be around somewhere; you just have to find them.

Now a little bit about prawns and shrimp. There is frequently great debate over seafood items called "shrimp," "prawns" and "scampi." (What you eat, by the way, is the tail meat, not the whole animal.) There are hundreds of different species of this group, and the common names generally reflect size. Small ones are known as shrimps; larger ones are generally referred to as scampi or prawns.

Did you know that the last time you savoured the buttery flesh of fresh crab or shrimp you were actually eating a relative of the common sow bug? Unlikely as it may seem, crabs, shrimps and their marine relatives share an ancient kinship with an enormous group of animals, including insects, spiders and centipedes. Collectively these are known as the phyllum Arthropoda, meaning "jointed limb."

The marine arthropod branch of this extensive group is known as the crustacea, or "crusty ones," and has a membership of over 26,000, including crabs, shrimps, lobsters, barnacles and related forms. Although some of these occur on land and in fresh water, the majority make their home in the sea.

CRABS

The higher crustaceans, such as crabs, can be thought of as a simple body with eighteen pairs of appendages. Unlike the related centipede, which has a whole series of identical legs, the crab has adapted pairs of appendages to various functions, such as chopping, walking and baby-sitting.

Beginning at the head end, there are two pairs of antennae, which serve as sensory organs of touch and perhaps smell. Along with the stalked eyes, balancing organs and sensory bristles, the antennae form the sensory equipment of the crab.

Next are a number of modified appendages known as mouth parts, some functioning as knives and forks—cutting, picking and sorting food. Other mouth parts pulverize and handle food, doing the same job as our teeth and tongues. Food thus prepared is pushed into the jawless mouth and passes directly to the stomach, where it is further ground by stomach "teeth," acting much like a food mill. From the stomach, food passes through the intestine, where nutrients are absorbed, and the unusable "wastes" are excreted through minute openings at the head end and through the anus at the rear.

After the mouth parts come the most universally recognized appendages, the pincers, or chelipeds. One of these is often larger than the other, particularly in males. The function of the pincers is grasping, tearing and defence. To avoid being nipped, pick up a crab from behind or from above, across the top of the shell.

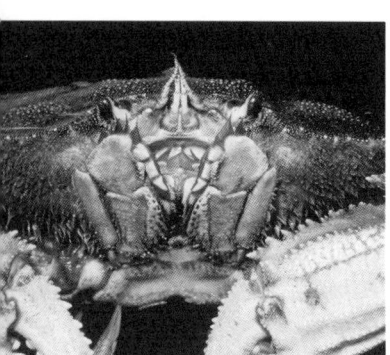

The edible or Dungeness crab (*Cancer magister*) ranges from California to Alaska from low tide level to 100 fathoms and is part of an important crab fishery. Legal size for the species is 16.25 centimetres (6½ inches) across the carapace. Females mate after molting during the summer in inshore waters and carry up to one million eggs until hatching the following spring. Large, edible crabs may be 25 centimetres (10 inches). They feed on clams, marine worms and even small fish. Look for these crabs in sandy or muddy bays where there is plenty of eelgrass.

Four pairs of walking legs follow the pincers. The legs on one side of the body pull while those on the opposing side push, causing the crab to move sideways. Attached to the base of each leg on the underside are the breathing gills.

One would think the appendage supply would by now have been exhausted; not so. The crab can be visualized as a shrimp with its tail tucked permanently beneath itself in a kind of upside-down apron. On the underside of this "apron" are a number of small legs. The female uses these to hold and brood her eggs. The shape of this uptucked tail is determined by the sex of the crab. In males it is narrow and triangular, whereas in females it is broad and U-shaped, the better to accommodate her large number of eggs.

Many crabs of the West Coast mate in the fall. The female holds the male's sperm in a special abdominal receptacle until the eggs are extruded and fertilized. She then holds the eggs under her tail from late autumn until they hatch and are released in the spring. Crabs hatch into a very uncrablike form—a minute swimming zoea. Several molts and some months pass until a miniature crab develops, ready to take up life on the sea bottom or shore.

Crabs grow by frequent shedding of the shell. Not only does this shell keep the outsides out and the insides in and serve as an effective protective armour, but it is a skeleton too, providing muscle attachment for an animal that has no bones. Once hardened, this exoskeleton, as it is called, cannot be added to or expanded. It is a prison and must be shed if the animal is to grow. This process of shedding the exoskeleton is known as molting.

In preparation for molting, hormonal activity stimulates the production of a new, soft, chitinous shell beneath the old. When ready, the crab shrinks somewhat and backs out of its old suit through a crack at the waist line, leaving behind every hair and hump, including the lining of its stomach and its stomach teeth. While still soft, the crab takes in water, expanding the new soft shell to a more desirable dimension, giving an overall increase of 11 to 29 per cent, depending on the age and, to a lesser degree, the sex of the crustacean. Younger animals show a proportionately greater shell increase per molt than older ones.

Shore crab (*Hemigrapsus nudus*), which has a smooth, square carapace, is a very common crab of the intertidal zone of rocky shores. A large individual measures about 3 centimetres (1¼ inches) across the back. Colouration in the species is quite remarkable, ranging from dark to light, and including mahogany, green, brown or white, often in fascinating combinations. The shore crab is a scavenger and ranges from Alaska to the Gulf of California.

Graceful kelp crab (*Pugettia gracilis*) inhabits seaweeds and is frequently camouflaged with settling organisms on its carapace, suggesting a more sedentary lifestyle than that of the very active northern kelp crab.

Northern kelp crab (*Pugettia producta*). Strong and aggressive, this species is most often found clinging to the brown seaweeds it mimics so well in body texture and colour. It is generally clean and free of settling organisms. It grows to 10 centimetres (4 inches) across the carapace. In the summer, this crab is a vegetarian, eating brown algae. In the winter, if plants are not available, it switches to a diet of barnacles, hydroids and bryozoans.

The actual molting time (that is, the time taken to shed the old shell) is only about fifteen minutes, but the newly molted crab is vulnerable to predators for at least forty-eight hours before the new shell hardens. Some crabs eat the recently molted "cast" to quickly supply their bodies with the great amount of lime salts needed to harden the shell. Still, many casts can be seen on the seashore, so perfectly intact they are often taken for dead crabs. The average number of molts, not counting larval molts, for the Pacific edible crab (*Cancer magister*) is in the neighbourhood of fifteen in its life span of five to six years.

In a process called autotomy, crabs are able to part company with their legs or pincers if it is in their best interests to do so—for example, to escape. A spasmodic contraction of the muscles near the joint between the leg and the body (an area known as fracture plane) releases the limb with a minimum of bleeding. Only blood vessels and nerves—no muscles—pass through this joint. A stub is rapidly formed, and over a series of molts, it regenerates a new, functional limb.

Throughout the world, crabs exhibit enormous variation in size—from the huge, 3.75-metre (12-foot) span of the Japanese spider crab (*Macrocheira kaempferi*) to the minute burrow crabs. They vary equally in habitat, from the tree-climbing coconut and land crabs of the South Pacific to the parasitic and deep-water marine forms.

On the West Coast, crabs are strictly marine, forming a large and active army of housekeepers scavenging the shore and sea floor for food. Some are masters of camouflage: the kelp crab (*Pugetti producta*) perfectly mimics, in texture and colour, the slick olive seaweed where it makes its home; the decorator crab (*Oregonia gracilis*) masks its presence by securing a collage of plants and animals to its carapace; and the hermit crabs (genus *Pagurus*) are entirely unrecognized until they move.

The hermit crab should not be assumed to be just any crab that chooses to protect itself by climbing into an empty snail shell. It belongs to a whole separate family of crabs—not quite crabs, yet not quite shrimp or lobsters either. This "crab" has a soft abdomen, and in some species it is flexed to the right to accommodate the spiraling of suitable snail shells. The tail has become modified to a hook for holding onto the shell house, and the large claws are shaped to block the shell's en-

This decorator crab (*Oregonia gracilis*) is shown without decoration. Spiderlike with small pincers, this crab is often so well covered with a luxuriant growth of seaweed and hydroids that it is invisible. The decorator does not simply tolerate or encourage settling organisms but actively decorates itself with them. When moved into a new environment, the decorator will redecorate itself to suit its new surroundings. The species grows to 5 centimetres (2 inches) across the carapace and ranges over rocky shores from California to the Bering Sea.

Decorator crab with decoration.

The black-clawed crab (*Lophopanopeus bellus*) looks somewhat like the shore crab but is heavier bodied and has heavier, black-tipped claws. Unlike the shore crab, the black-tipped crab does not generally run when handled but rears up its hind legs and becomes rigid. It is found intertidally under rocks in muddy sand. The carapace measures 2.5 centimetres (1 inch) across. The species ranges from Alaska to Baja California.

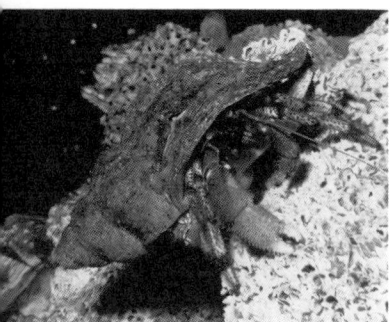

Hermit crab (*Pagurus sp.*) can be found in all kinds of snail shells, though some species show a marked preference for a particular kind of shell. Crabs will take shells from each other but will leave the shell alone if it is still occupied by the snail. Hermit crabs are great scavengers of plant and dead animal material. Tide pools are a good place to look for them.

Pea crabs (*Pinnotherid crabs*). A number of large West Coast clams and mussels are host to small crabs that live in the mantle cavities of the host bivalves. Small, less than 2 centimetres (¾ inch) in diameter, and soft bodied, they appear to feed on mucus and suspended food particles brought into the clam with the water it circulates for its own food and respiratory needs. Male pea crabs are typically much smaller and more mobile than females. Clams found hosting pea crabs are still edible.

The porcelain crab (*Petrolisthes eriomerus*) is a small, flattened crab, generally 2 centimetres (¾ inch) across the carapace. It has only four pairs of walking legs, as opposed to the "true" crabs, which have five. This fact indicates that the porcelain crab shares kinship with hermit crabs, which also have four pairs of walking legs. (The fifth pair is tucked up inside the body and is not functional as walking legs.) The porcelain crab is most common to exposed rocky beaches under loose rocks and ranges from British Columbia to southern California. The porcelain crab is a filter feeder.

trance when the animal is withdrawn. Like other crabs, the hermit crab must molt to grow, and it must find increasingly larger shell houses to complement each new body size. Hermit crabs reproduce in June and July on the West Coast, the females brooding their eggs in the same manner as other crabs.

Another fascinating little crab has surrendered the security of its own hard shell for another home. It is the parasitic pea crab *(Fabia subquadrata)*, the female of which is commonly found in the mantle cavity of the California mussel *(Mytilus californianus)* and the horse clam *(Tresus capax)*. Once established, the crab lives a life of safety and comfort. The male of the species is much smaller and is free living. He pays occasional visits to the female for purposes of reproduction, being small enough to enter and exit between the mussel's shells. Most females become so large and soft that they are never able to leave their hosts.

SHRIMPS AND LOBSTERS

No native lobsters occur on the West Coast north of California, and attempts by the Canadian Ministry of Fisheries to introduce the Atlantic species in British Columbia have been unsuccessful as a commercial proposition. Some Vancouver-based divers claim a good number of lobsters are alive and well, but they are not saying where the potential gourmet dish is.

In contrast to the dearth of lobster is the abundance of shrimp species. Of the eighty different kinds of shrimps identified in British Columbia, six species are harvested commercially for the tail meat. This is the strong muscle used by the shrimp to propel itself backward. By spreading the end of the tail like a Chinese fan and quickly pulling the tail forward, the shrimp is able to beat a hasty retreat. Forward movement is achieved by walking or jumping.

Generally, the basic physiology, growth and reproduction in shrimp parallels that of the crabs, with one interesting exception. It has been found that *Pandalus* (coon-striped shrimp and prawn) shrimps mature and function first as males, then at 2½ to 3 years of age they pass through a transition, or intersexual phase, to become females until their death at 4 to 5 years of age.

Many crabs, lobsters and shrimps engage in some sort of grooming activity, ranging from a simple scraping in

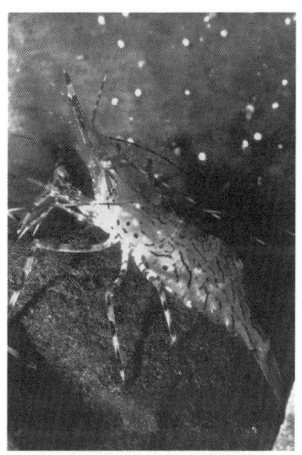

Coon-striped shrimp (*Pandalus danae*). Sand or gravel bottoms where a rapid current exists is the favoured habitat of the coon-striped shrimp. A large specimen may be 12.5 centimetres (5 inches). The species ranges from Alaska to California and is harvested commercially.

Pacific prawn (*Pandalus platyceros*) is the largest of the commercial shrimps, growing to 22.5 centimetres (9 inches). It generally occurs over rocky bottoms from Alaska to San Diego. Like some other shrimps, it functions first as a male in its second year, changing sex to become female during the third and fourth years. The Pacific prawn is a commercially valuable species.

Ghost shrimp (*Callianassa sp.*) lives in a U-shaped burrow, which it digs using its mouth parts, carrying dirt and sand to the burrow's opening with its legs and claws. Food consists of organic material extracted from the mud or plankton from water moving through the burrow. Holes in muddy sand at low tide in bays and estuaries may be the home of this animal, not clams, as is usually thought.

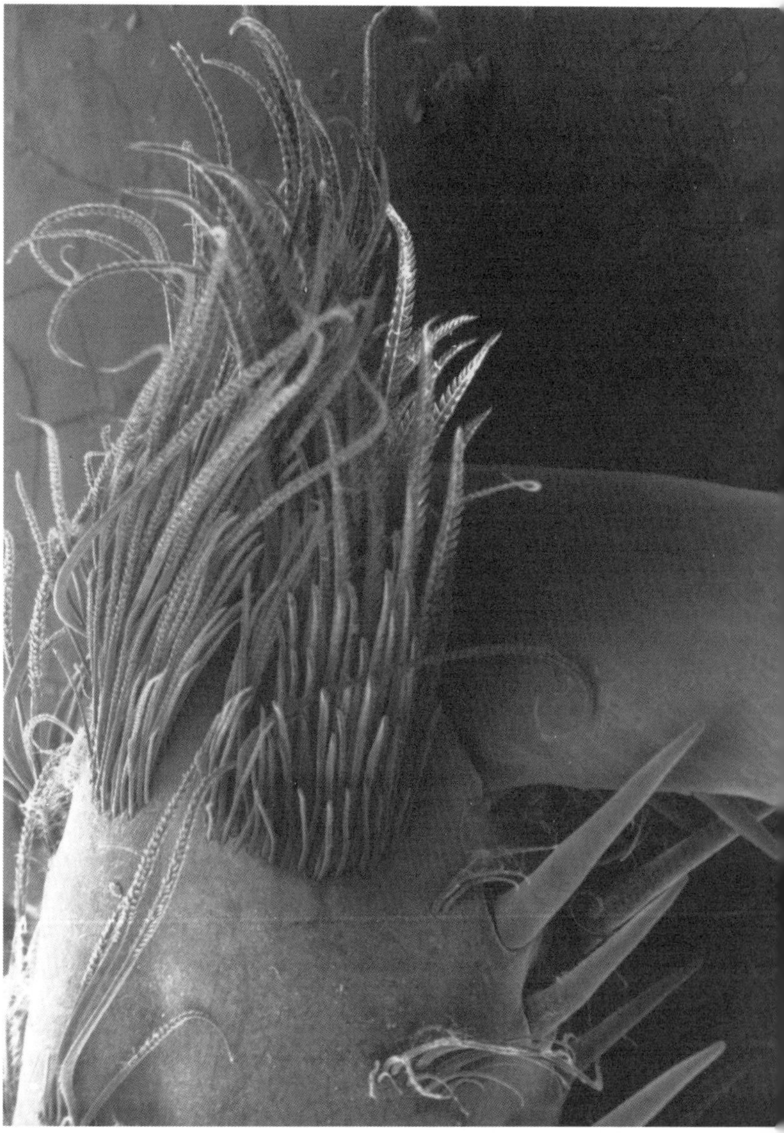

The last walking leg of the coon-striped shrimp (*Pandalus danae*) magnified fifty times. The grooming brushes (setae) on the legs are used to keep the shrimp's body surfaces and gills free of debris and settling organisms, including parasites. It is believed that the ability of the shrimps to "clean" themselves has had much to do with the success of the decapod (ten-legged) crustaceans and their radiation or branching into so many different groups.

crabs to remove unwanted settlers, to a highly refined process in a group of shrimps known as carideans. Antennae, appendages, gills and general body surfaces are brushed and cleaned with special clumps of setae (bristles), which form brushes on some of the appendages. It has been established experimentally that if the grooming brushes are removed and the shrimp is, therefore, unable to preen, the animal's gills become so fouled with sediment and detritus that the shrimp dies of asphyxiation. This anti-fouling activity, not just of the gills but of the entire body surface, effectively prevents the settlement of sedentary organisms, including parasites.

ISOPODS

Many small species of buglike crustaceans occur in the marine environment under rocks, on floats, clinging to seaweed or burrowing in wood. These are known as isopods, a term meaning "equal foot." They range in size from a few millimetres to 5 centimetres (a fraction of an inch to 2 inches). They have no pinching claws, as crabs do, but generally cling with all eight claw-tipped legs. Over four thousand species have been recorded, a number of which are parasitic on shrimp, lobsters and crabs. The 2-millimetre common gribbles are responsible for the fine bore holes seen on the surfaces of much beach wood, floats and pilings. Gribble burrows are easily distinguished from those made by the teredo, or shipworm (a species of burrowing clams) by their much smaller size.

AMPHIPODS

Another large group of buglike crustaceans common on wharves, floats, seaweed and the beach is the amphipods. The group has in the neighbourhood of 3500 species and includes the beach hoppers and sand fleas. Amphipods are typically flattened from side to side, as opposed to the isopods, which are typically flattened from top to bottom. Most amphipods are small, under a centimetre (½ inch), have proportionately long antennae and antennules, and appear to be hunched over.

Isopods. These small, buglike crustaceans range in size from a few millimetres to 5 centimetres (a fraction of an inch to 2 inches). They cling to rocks, floats and seaweed with claw-tipped legs. Isopods are secretive, hiding just about anywhere. Many species are nocturnal, emerging at night to forage.

Amphipods. This large group of small, buglike crustaceans has long antennae and the appearance of being hunched over. They are distinguished from isopods by being flattened from side to side. Amphipods are abundant and diverse. One species lives exclusively on whales as whale lice.

BARNACLES

Next time you find yourself climbing over rocks near the shore, cursing the plague of barnacles cutting and chewing painfully at your feet, get down on your hands and knees and meet this infamous creature eye to eye.

Barnacles are fascinating little animals, incapable of fleeing their enemies. Yet they eat, respire, sense the world around them and reproduce. In seas the world over, 750 kinds are found, at least 23 species on the West Coast.

There are two basic variations: the acorn barnacle and the gooseneck barnacle. The acorn barnacle is well known as a small, white, firmly attached volcano-shaped structure. Its shell is usually four rigid and fused upright plates with two additional pairs of movable plates at the summit. These movable plates function as doors, sealing the animal within against the ravages of exposure and predation. The gooseneck barnacle is functionally and anatomically similar, but it looks different; its shelled body is attached by a fleshy stalk, or neck, making up two-thirds or more the length of the complete animal. This barnacle, homely in the extreme, looks like a bird's beak on the end of a neck. In fact, stories of birds and barnacles being related abound in old accounts of natural history.

One such account, written in 1597 by the naturalist Gerard, stated that barnacles grew on certain seaside trees in the Orkney Islands, north of Scotland. At maturity, so the narrative continued, the shells would open,

Cross-section of an acorn-type barnacle.

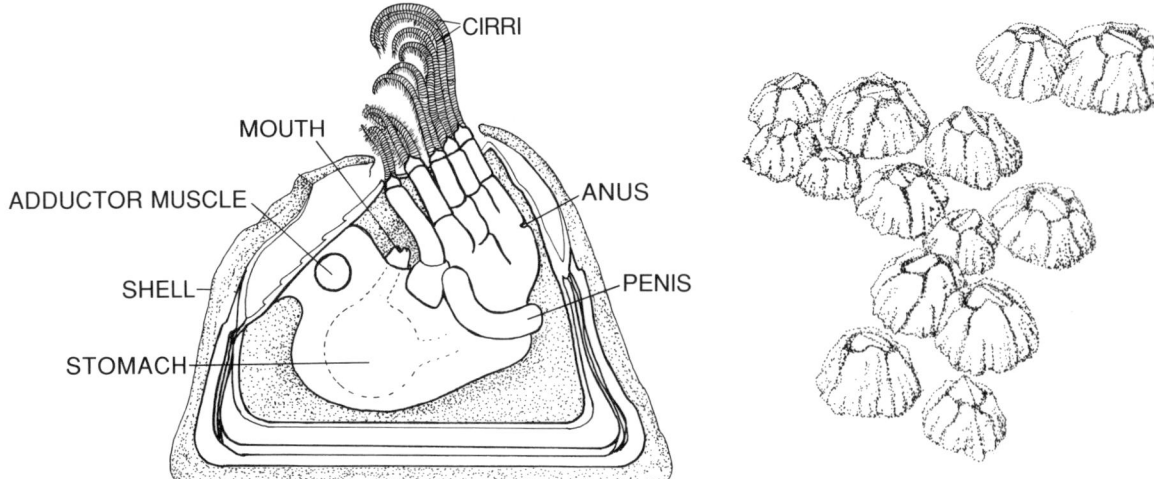

CIRRI

MOUTH

ADDUCTOR MUSCLE

ANUS

SHELL

PENIS

STOMACH

giving rise to a species of goose called barnacles. Infertile barnacles fell to the ground and died. Other similar accounts exist, most probably growing out of a confusion of language and alteration of names. In addition, the arctic breeding grounds of the barnacle goose were unknown at the time, and the "barnacle tree" hypothesis accounted for the existence of progeny otherwise unexplained.

Still later, barnacles were erroneously classified with clams and snails. It was not until J. Vaughan Thompson observed the free-swimming larval form of the barnacle in 1830 that it was established as a crustacean in the same company as crabs, lobsters and sand fleas.

The young barnacle is a free-swimming form known as a nauplius. At this stage of its life it closely resembles its juvenile crustacean relatives; it is minute, with a single eye, three pairs of appendages and some means of flotation, usually a drop of oil. This ensures that the tiny animal will float near the water's surface, where its food, the phytoplankton, is abundant. The barnacle eats ravenously, growing rapidly and molting every three to five days until it has passed through seven stages. The juvenile now has a pair of hinged shells and is called a cypris. It is this cypris that settles down to the real business of being a barnacle. At only 25 millimetres (¹⁄₁₀₀ inch) long, it searches for a home site and, once a site is found, glues itself to the spot with a secretion from adhesive glands found near its head. Other glands ooze shell-building material, and in no time it has begun to look and behave like an adult barnacle.

Since the barnacle is to be sedentary for the rest of its life, from one to seven years, it no longer requires its six pairs of legs for walking. Instead of discarding them, the barnacle brings them into service as food catchers. The curled, feathery legs extend through the open plates at the barnacle's summit and begin sweeping the water for particles of food—plankton and detritus. Other appendages near the mouth bundle the entrapped food and convey it to the mouth, where it passes to the stomach for digestion. This peculiar feeding mechanism led Huxley to describe the barnacle as "a crustacean fixed by its head kicking food into its mouth with its legs."

The barnacle is without a true heart or blood vessels, so nutrients from digested food, and oxygen absorbed

Nauplius larva.

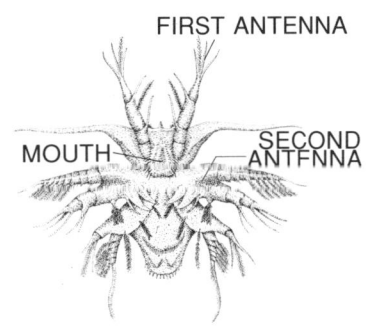

FIRST ANTENNA

MOUTH

SECOND ANTENNA

Acorn-type barnacles. Note that the movable plates are shut tight to protect the animal from drying out when exposed at low tide.

A barnacle with its cirri extending out the top of its shell. The cirri sweep the water for tiny particles of floating food.

from the water, are distributed to muscles and various organs by fluid flowing through passages between them, a simple but efficient method of circulation known as lacunar circulation.

But how can an imprisoned animal like the barnacle grow? The barnacle partially replaces its shell. It is now thought that the exterior plates grow and are added to, whereas the internal support is shed as a new shell, developing beneath, is expanded with water to allow for greater size. It then hardens rapidly. Molts can be very rapid, only days apart, depending upon species, temperature, food and breeding cycles. Whole beds of barnacles may molt simultaneously.

Barnacles may also breed simultaneously with a reproductive potential that is awe-inspiring. One and a half kilometres (1 mile) of shore can have more than 1.5 billion barnacles bringing forth 1500 billion potential young! Barnacles are hermaphrodites, having both male and female sex organs, but they usually do not fertilize their own eggs. Fertilization is achieved, in at least some species, by means of a long, extensile penis, which transfers sperm to the mantle cavity of a neighbouring barnacle. After fertilization, the many thousands of eggs are brooded for perhaps two weeks before the hatched larvae are released. It is a small wonder, then, that a ship can become heavily fouled with barnacles in as little as eight months. On a large ship this fouling can amount to tons of additional weight and a drastic reduction in speed, with proportional increases in fuel consumption and maintenance costs. Since at least the fifth century B.C. (when the first records of this problem were written) seagoing people have been trying to frustrate the settling attempts of barnacles. In modern times, toxic paint, copper sheathing, chlorine gas, radioactive paint, heat and electric impulses have been tried. Researchers continue to explore the effects of toxins on barnacles in an effort to find a lasting solution to this expensive problem.

Other researchers are looking to the barnacle for other reasons. Fossil barnacles that attached themselves to shells 15 to 20 million years ago are still attached, testimony to their adhesive capacity. What fantastic potential there is in this substance that lasts 15 million years, provides a sheer strength of 454 kilograms per 6.45 square centimetres (1000 pounds per square inch), will bond tissue and bone, can be heated to 177° C (662° F) and not melt or frozen to −146° C (−383° F) and not crack or peel, and is not attacked by acids, alkalines or organic solvents. The applications that a successful synthesis of this barnacle glue can give to industry, medicine and dentistry are endless.

A FASCINATING VARIATION

Coronula, a whale barnacle, is one of a few species that attach themselves to the skin of certain whales. Another is said to grow only on the tongue of a particular turtle. But the worst freeloaders of all are the sacculinids, which are truly parasitic barnacles, unlike the former, which are simply inconveniences. The larval *Sacculina* reaches its victim while in the free-swimming stage, attaching itself to one of the hollow bristles on the crab's exoskeleton. The bristle is pierced, and a few parasitic barnacle cells are released into the body of the crab. These cells come to rest at the junction of the stomach and intestine, where they grow, extending rootlike structures throughout the crab's body. The *Sacculina* barnacle now feeds on the crab, destroying its reproductive organs and altering the crab's sex hormones so that regardless of the crab's previous sex, it molts to become a female, complete with an apronlike abdomen. During the molt, the parasite is not shed but remains, extending a saclike mass through the abdomen to the outside, while the host's shell is still soft. The sac contains the self-fertilized eggs of the *Sacculina*, which are soon released to become more parasitic barnacles. Parasitism by a *Sacculina* barnacle does not necessarily spell the end for a victimized crab.

Gooseneck barnacle (*Pollicipes polymerus*) clumped among California mussels (*Mytilus californianus*). Where a number of individuals occur, all will be oriented in the same direction so that as waves fall back off the rocks, the spread, feeding cirri of the barnacle will capture floating organisms and small animals. Individuals of this species grow to 25 centimetres (10 inches) and occur near the mid to low tide level on exposed coasts.

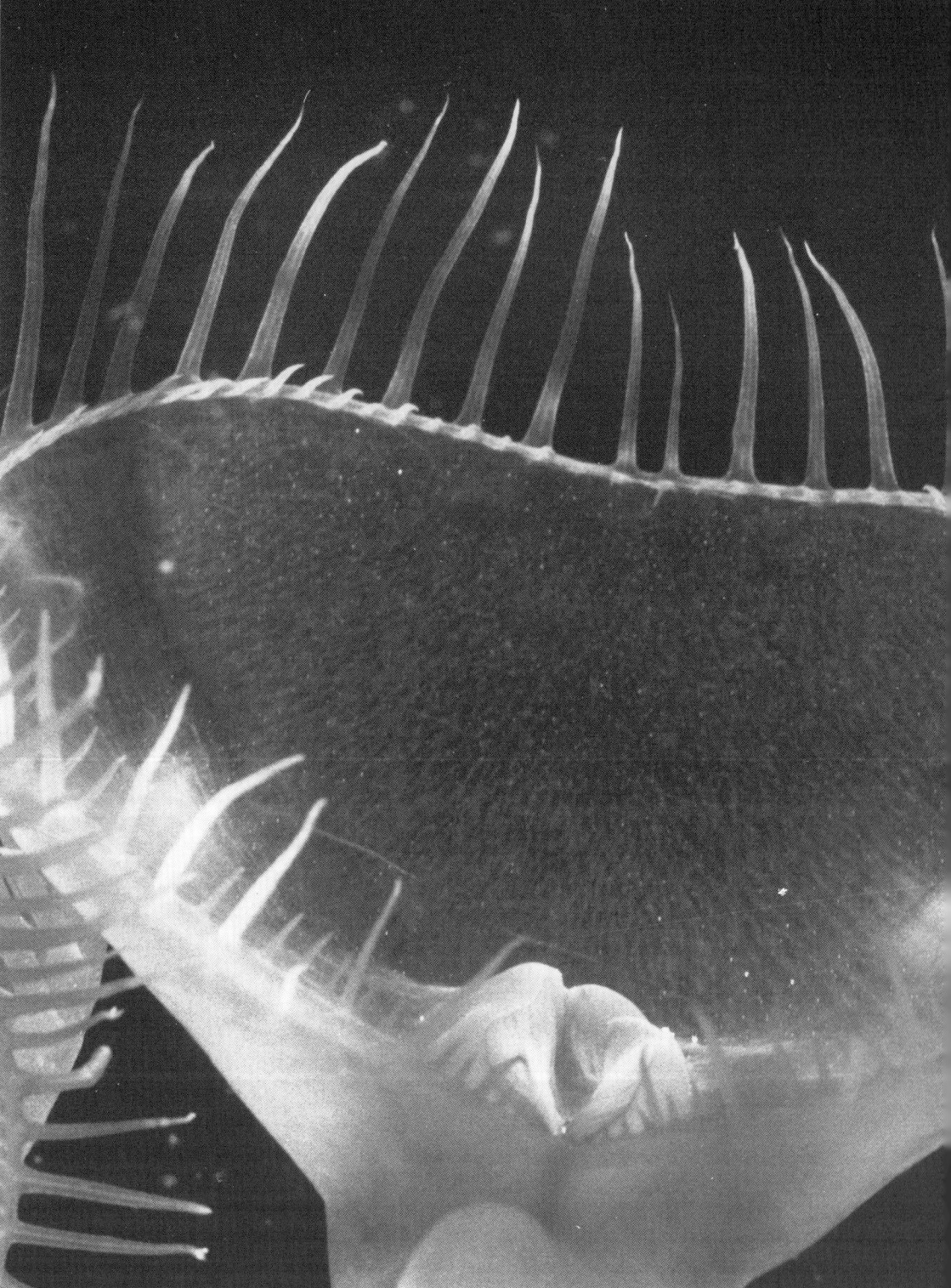

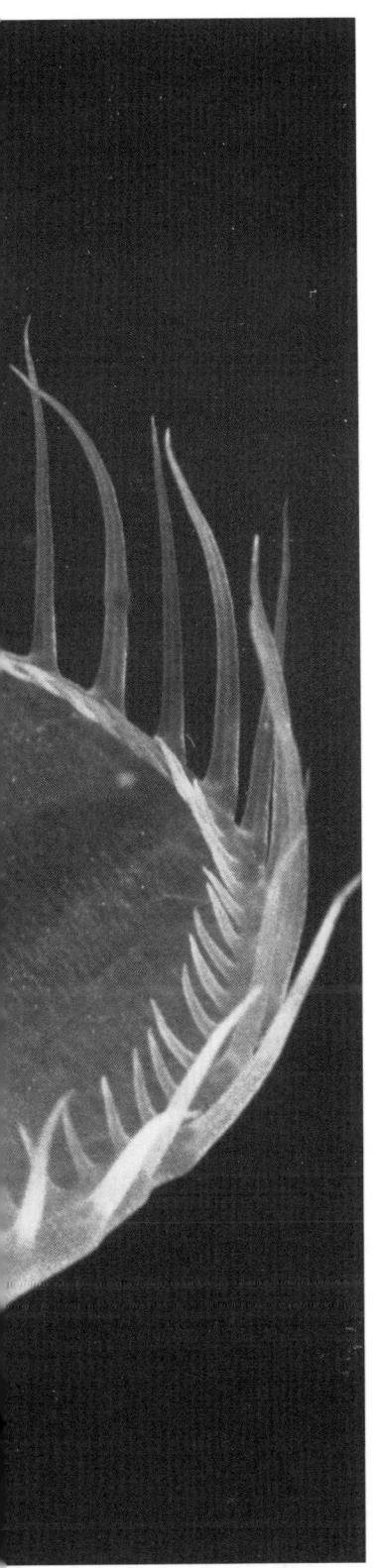

CHAPTER 7

Mollusks

There seems to be a rule that goes something like this: the more protective armour an animal has, the slower it is. This makes a certain amount of sense, since the more protection an animal has, the less reason it has to run from its enemies.

The mollusks are a group of animals that have exchanged speed for security. These are the animals that produce the objects we know as seashells. A seashell is in fact an animal's house or armour. Single shells like the snail shells, conch shells, high-hat shells and abalone shells are all variations on a theme. A sluglike animal secretes a shell that is added to as the animal grows, and it usually coils as it is added to. Double-shell animals such as clams, oysters, scallops and mussels wear a pair of shells, like an oversized saddle. The shells join on the animal's back, one hanging down on each side, with the animal in between.

Some mollusks used to have shells in the distant past but gave them up, opting for speed over security. A slug is really just a snail without a shell. Octopuses and squid not only gave up their shells but also developed arms out of what were feet and took to the open sea by using jet propulsion. Before fish came along, octopuses and squid ruled the oceans because they could swim (at least that is the theory). Fish subsequently out-competed the octopods, and their heyday was over. Octopuses and squid are still around, but they are not the dominant forms they probably once were.

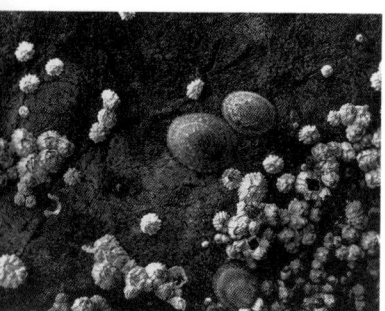

Limpets (*Notoacmea sp.*) grow to about 6 centimetres (2¼ inches) and feed by grazing microscopic algae growing on a rock. Their defence against predators is to clamp down firmly.

Keyhole limpet (*Diodora aspera*) has the same Chinese-hat shape as a regular limpet and, in addition, an opening at its summit. This species is found intertidally from Alaska to northern Baja California and grows to 5 centimetres (2 inches). A scale worm is frequently found living on the limpet's foot. When the limpet is attacked by a purple sea star, the worm bites the sea star's tube feet, causing the sea star to withdraw.

ABOUT MOLLUSKS

Since prehistory, humankind has eaten mollusks, used them as tools and later as ornaments and currency, and even worshipped the shelled creatures. They were abundant, accessible and often exquisitely beautiful.

This animal group is of such spectacular diversity that there seems to be no beginning and no end. How else does one describe a group of living forms that includes the garden slug, shipworm, clam, snail, abalone and octopus? All are mollusks, and they number some 110,000 living species, with many more fossil forms. Their history dates back at least 500 million years.

The mollusks have invaded every habitat short of taking to the air. They can be found at depths where hydrostatic pressure measures 563 kilograms per square centimetre (4 tons per square inch), on the snow-capped Himalayas, on the desert and in fresh, salt and brackish water.

The grand group of mollusks has been broken into seven smaller classes based on shared characteristics and similarities. The commonest of these are (1) the Polyplacophora, having eight shells, as in the chitons; (2) the Scaphopoda, having long, tubular shells, as in the tusk shells; (3) the Pelecypoda, having two usually equal-sized shells, as in the clams; (4) the Gastropoda, having one shell, as in the snail, or none, as in the slug; (5) the Cephalopoda, having tentacles and a shell, as in the nautilus, or having no shell, as in the octopus; and (6) the Monoplacophora. The last class was previously thought to have been extinct for 450 million years, until a living specimen was dredged from a deep trough off the coast of Costa Rica and described in 1957.

All molluskans, though fantastically diverse in habit, form, size and habitat, share among themselves, and with no other animal group, two characteristics. These are the mantle and the radula, though not all mollusks have a radula. The mantle is a fold of soft flesh enclosing the gut, or viscera, and is responsible for secreting a calcareous shell. The radula is a toothed tongue. Any other generalized statement about mollusks would demand qualification. Therefore, information of a more specific nature will be provided as each molluskan group is described in the following sections.

All classes, except the Monoplacophora, have representative species on the West Coast. The area claims to have in excess of 900 marine shells. This figure, of course, excludes the non-shell-bearing marine mollusks such as the nudibranch and octopus.

In conclusion, it must be remembered that the shell, though invaluable for identification, is not an animal. It is housing and protection; the animal lives within. Scientists studying mollusks recognize this vital difference and label themselves accordingly: a malacologist deals with the animal (soft parts), and a conchologist deals with the shell (hard parts).

I invite you to become both. An excellent place to start is on the beach. Examine the bits of shell and debris left at the drift line for clues as to which living mollusks may be found there.

SNAILS AND SEA SLUGS (GASTROPODS)

Have you ever wondered why a snail's shell is coiled? Imagine that you are a snail, carrying your house on your back, and that you must add to your house as you grow. With time, the shell would become higher and higher and soon being to topple and fall. Your house would no longer be an asset but a liability. Consider how much more practical it would be if the house turned in on itself in a low spiral, creating an expandable yet compact and portable abode.

All modern snails and their relations, including slugs, are collectively known as gastropods. All have a spiraled shell, at least at their beginning, even if the spiraling may not be obvious, as in the Chinese-hat snails and the limpets, or if the shell is reduced or present only in very early stages, as in the slugs.

Where present, the shell is carried on the snail's back. The top of the spiral, or apex, is usually directed to the hind end of the animal, with the shell opening, or aperture, on the creature's back. The apex is held in place by a strong muscle attached to the columnella, or central axis of the shell. In addition to the large body of the main shell, there may be an operculum, or door plate, on the rear top of the snail's foot. This is held against the aperture when the animal withdraws into the privacy, security and protection of its shell. Less highly evolved snails, such as the abalone and limpet, have no operculum and must rely on the tenacity of their grip for protection. Since it is estimated that the abalone may exert a suction equal to 4000 times its own weight,

Abalone (*Haliotis kamts-chatkana*). At least 70 species of abalone occur throughout the world. The northern abalone occurs from Alaska to Point Conception, from the low intertidal to 18 metres (60 feet) in depth. Abalone are grazers, feeding on the algae and diatoms encrusting the rocks where they live, and are distinguished by a row of holes along one side of the large, flat, ear-shaped shell. The northern abalone grows to 15 centimetres (6 inches) in length.

Abalone are taken commercially for the restaurant and supermarket trade and recreationally are subject to size and bag limits. People have been major predators on abalone for at least seven thousand years, as indicated by shell mounds on the California coast.

this gripping ability is a good alternative.

Gastropods, also known as univalves, meaning "one shell," are only one branch of the huge mollusk group, which includes chitons, clams and octopuses. The gastropods themselves have become incredibly diverse, not only in appearance but also in size, habits and habitat. This great diversity has resulted in many fascinating adaptations to the various environments they inhabit and the food sources they exploit. Consequently, few generalizations apply to all gastropods.

Gastropods are so named because they seem to creep on their bellies. The name is derived from the Greek *gaster*, "stomach," and *podos*, "foot." Generally, a sheet of mucus is secreted by the forward portion of the foot so that the snail glides along on its own slime, using the rippling muscle waves of the foot's inner sole. In some gastropods the foot is modified to float on a bed of tiny moving hairs. In others, such as the local chink shell (*Lacuna variegata*), one half of the foot moves and then the other, in much the same manner as a person whose ankles are tied together. The result is a "waddling snail." Still other gastropods, like the bubble shell (*Haminoea virescens*), have extended the sides of the foot into wing-like flaps for swimming.

In general gastropods must be able to move about, however slowly, in order to eat. The gastropods must seek out their sustenance, whether they are grazers, herbivores, scavengers or predators. Whatever the food preference, the gathering device is generally the same— the radula. Best described as a long, tooth-bearing tongue, the radula is held in a tubular snout, or proboscis, when not extended and being used for scraping algae or flesh. Saliva and digestive glands act on the eaten food as it passes through the digestive tract. Unusable products are eliminated via the anus into the mantle cavity, which is just above the head in most marine snails.

The majority of marine snails are grazers or nibblers, eating whatever plants and small animals that can be scraped from the rocks. Others are scavengers, eating bits of detritus and dead organisms, and in so doing they fulfill a most important function—a combination of housekeeping and recycling. Periwinkles, or littorines, are very common little snails of the West Coast intertidal areas, and they have reached the ultimate in

Dogwinkle (*Nucella canaliculata*) grows to 2.5 centimetres (1 inch) and is common to rocky shores. It ranges from Mexico to Alaska and has a wide colour range. The dogwinkle feeds on barnacles and mussels. The snail spends one to two days drilling and consuming a prey item.

recycling by actually feeding on organic matter contained in Eocene and Cretaceous silt stone as they rasp and scrape the encrusted rocks. It has been calculated that in 2.6 square kilometres (1 square mile) of beach in La Jolla, California, there may be 860 million littorines eroding more than 2000 metric tons (about 1800 tons) of this stone in a year.

Other snails are not content to scrape algae slime or to eat food that is 100 million years old but prefer live prey. Bulbous moon snails *(Polinices lewisii)* plow through sand and mud, stalking defenceless clams; Thais snails creep with singular purpose over barnacle beds, dispatching them as they go; and oyster drills attack succulent young oysters.

There would seem to be no contest in such a chase, when one considers how difficult it is to part the shells of a clam or an oyster that chooses not to have its shell parted. What chance does a handless snail have? Carnivorous snails make no attempt to compete with the tremendously strong adductor muscles of the bivalves, which keep the shells so tightly closed; instead, the predatory snail breaks through the defence of its victims by using a combination of a shell-softening secretion and a drilling radula. Alternately softening the shell and scraping it away, a process that may take many hours, the snail eventually gains entrance to the shell's interior. Using its proboscis like a soda straw, the snail then sucks out the flesh, and dinner is served.

The precise drill holes frequently seen on the umbo (hinge) region of clam shells cast up on a sandy beach are a clear indication that moon snails are about, hunting down clams.

A pair or two of sensory tentacles, a pair of stalked eyes, and an assortment of sensory cells provide the gastropod with information about its environment. In some species, the eyes may be sophisticated organs having a lens and retina. However, most sensory messages received will be chemical ones—subtle changes in the composition of the water around the snail. An example of the efficiency of this system is demonstrated by the fact that an oyster drill will choose a young, still-growing oyster over a mature oyster on the basis of metabolites specific to shell production secreted by the younger oyster.

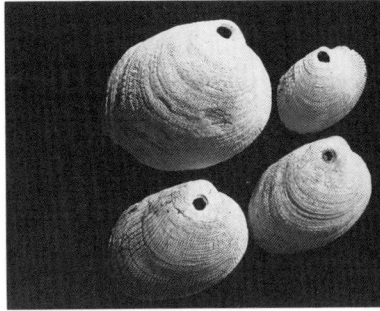

Clam shells that have been bored by the proboscis of the moon snail typically bear a countersunk hole in the umbo region. Once entrance to the clam shell has been gained, the moon snail inserts its proboscis and eats the clam.

Thais snail (*Nucella lamellosa*) has a very variable shell. Sculpturing ranges from smooth to very wrinkled, and colour ranges from white to orange-brown with a great variety of striping. This species grows to 3 to 7.5 centimetres (1¼ to 3 inches) and ranges from California to Alaska from the intertidal to 120 metres (400 feet). The thais are carnivorous snails, feeding on ascidians, clams, oysters and barnacles.

Also shown are the eggs of this snail, called sea oats. Time to hatching is nine to twenty weeks, depending on the temperature.

The egg case or "sand collar" of the moon snail (*Polinices lewisii*) does not encircle the eggs as in a nest but contains a layer of eggs sandwiched between two layers of sand. The whole is held together by a mucous section. In midsummer the eggs hatch into free-swimming larvae, which later settle and develop into the form of the adult.

The snail's chemoreceptors (sensory organs) must, of necessity, play an important role in finding a mate of the opposite sex and of like species in a crowded underwater world. Vision is the least useful for this purpose, since most snails do not possess eyes of sufficient complexity to recognize a mate.

Unlike the bivalves, most gastropods do not release eggs and sperm directly to the water to meet by chance; they mate and copulate. The male passes his sperm to the female via a penis located just behind and beside the head. The female then lays eggs in gelatinous capsules, with a number of larvae per capsule. Because of their distinctive appearance, the egg capsules of the Oregon triton *(Fusitriton oregonensis)* are commonly referred to as sea corn. Other marine snails and nudibranchs lay their eggs in a great variety of strings, coils and threads as temporary cradles for their progeny.

An unusual and frequently seen egg mass is the sand collar of the moon snail. The rubberlike collar does not encircle the moon snail's eggs like a nest, as one would think, but contains the eggs within the solidified sand-and-mucus collar itself. As the larvae develop, the collar begins to disintegrate, releasing the juvenile snails.

Contrasting sharply with the heavily laden and ponderous shelled snail, creeping its way among skittering crabs and swiftly moving fishes, is the graceful elegance of the gliding nudibranch, or sea slug. The richness of the nudibranch's colour and ornamentation appear at odds with a name like "sea slug," but it is the marine equivalent of a land slug. Both animals are considered true snails, though they do not sport shells. The larval shell has been completely discarded by the nudibranch but retained internally in the land slug.

Nudibranchs, for all their apparent lack of protective mechanism, are the forbidden fruit of many potential predators, looking delectable and tasting quite awful. Almost nothing will eat them, and they are left in peace to graze on sponges and hydroids. At least one, the diaphanous-hooded nudibranch *(Melibe leonina)*, uses its large, umbrellalike hood, fringed around its edges with feather-fine tentacles, to scoop suspended food particles from the water. If there is no current to deliver food into the hood, the nudibranch throws its hood over a prospective food item, closing over it in

Oregon triton (*Fusitriton oregonenis*). The living Oregon triton's shell is covered by a hairy periostracum. It is a large snail, growing to 10 centimetres (4 inches), and ranges from Japan to Alaska to Baja California, from the low intertidal region to 50 fathoms. The Oregon triton is carnivorous and known to feed on sea urchins. Eggs of this species are termed "sea corn" because the eggs look like kernels of corn.

The olive snail (*Olivella biplicata*), looks like an olive and is about the same size as the olive fruit. It is distinguished by a glossy and smooth exterior. This snail is found intertidally from Alaska to California. During the day it lives just below the sand. A carnivorous scavenger, it comes up around dusk and moves actively on the sand's surface at night. The snail buries itself again before dawn. Octopus, sea stars and the moon snail are predators on this snail.

A snail of the high intertidal zone, *Batillaria attramentaria* was introduced with oyster spat from Japan. It prefers muddy and sandy areas. This hardy snail can tolerate temperatures of 1° or 2° C (34° or 35.5° F), a wide range of salinities and up to sixteen days without water. It averages 3.5 centimetres (1¼ inches) and may live up to ten years. The small, round snails also shown are periwinkles (*Littorina sp.*). They inhabit the upper intertidal zone, often in tremendous numbers, where they graze on film growing on the rocks, or even on the rocks themselves. West Coast species are seldom larger than 1 to 2 centimetres (½ to ¾ inch) in length.

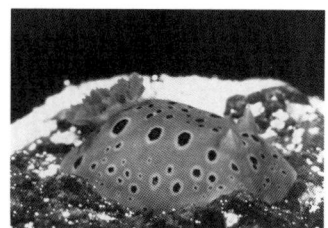

Leopard nudibranch (*Discodoris sandiegensis*) ranges from Alaska to California in rocky areas from the intertidal to depths of 40 metres (130 feet). Its colour ranges from white to dark chocolate, the former most often found subtidally, the latter intertidally. Leopard nudibranchs grow to 8 centimetres (3 inches) and feed on various sponges, including the bread crumb sponge. Look for this species on floating docks and in the rocky intertidal.

"Dorid" is the term given to oval, usually plump-bodied nudibranchs. There are two tentacles, sometimes retractile, on the head end, known as rhinophores. A circlet of gills surrounds the anal pore on the back towards the rear. The gills are known as branchiae and may be retractile.

Sea lemon (*Archidoris montereyensis*) is commonly found intertidally in rocky areas, in tide pools or near floats throughout its range from Vancouver Island to southern California. Five centimetres (2 inches) is the average length of this species, yet it may grow larger. Black peppering on a yellow ground and on the tubercles distinguishes the sea lemon from other nudibranchs. The plumed gills indicate the rear of the nudibranch; the stubby, paired tentacles (rhinophores) are at the head end. This species feeds on sponges.

Moon snail shell (*Polinices lewisii*). No other snail of the West Coast will be easily confused with this huge snail. Creeping over the surface or plowing beneath the sand on its huge foot, the moon snail searches for clams and cockles, eating the meat by drilling a neat hole in the umbo (hinge) region of the victim.

The moon snail shell grows to 13 centimetres (5 inches) in diameter. The snail ranges from Alaska to Mexico, from the intertidal zone to 46 centimetres (150 feet), and is preyed on by its own kind and the sunflower starfish (*Pycnopodia helianthoides*).

Opalescent nudibranch (*Hermissenda crassicornis*) is the most nearly ubiquitous nudibranch of the West Coast and can be found in eelgrass beds or in the rocky intertidal zone searching for its food: hydroids, tunicates, mollusks, eggs of various kinds and even bits of fish. The species ranges from British Columbia to California and grows to 5 centimetres (2 inches) in length.

72

the same manner as a South Sea Islander uses a cast net to capture fish.

BIVALVES

Single-shelled mollusks have great esthetic appeal in the graceful spiral of their often-elaborate shells. The appeal of the two-shelled mollusks, however, is largely gastronomic. No other food quite compares with the delicate flavour and smooth, boneless texture of a fresh, raw oyster seasoned with a drop of lemon and a dash of pepper. The eighteenth-century wit who claimed that "the man that will eat an oyster will eat anything" surely based his assumption on the questionable beauty of this creature when divested of its shelly fortress, rather than on the delightful flavour of the animal's flesh.

Oysters, with clams, mussels and scallops, form a group of some twenty thousand species. A few occur in fresh water, the majority live in the sea, and none exist on land. They are known as bivalves: *bi* meaning "two," and *valve* meaning "shell." The paired shells of the bivalve are generally, though not always, of equal size. They provide both protection and skeletal support for the animal held between them. In many species the shells are covered by a kind of dry skin known as the periostracum. A "hinge" joins the paired valves behind the umbo, or the bulged part of the shell, and is the back of the animal. To prevent the valves from falling open, strong muscles called adductors are firmly attached to both inner surfaces of the shells. Mussels and clams have paired adductor muscles, whereas oysters and scallops have but a single one. It is the large adductor muscle of the scallop that is sold as "scallops."

The term "pelecypod" is used alternately with the name "bivalve" for the same group of animals. It means hatchet-foot and refers to the large, muscular, wedge-shaped foot of the digging bivalves, such as the clams and cockles. Living in the mud, sand or gravel as they do, the clams and cockles must be able to move, if only to right themselves. To do this they extend their blade-like foot from between the shells to its full length, enlarge its end with blood so that it acts as an anchor, and then, by contracting the muscles of the foot, pull themselves toward the anchored end. The speed at which this operation is accomplished varies among the different kinds of clams. The razor clam *(Siliqua patula)* flies

Hooded nudibranch (*Melibe leonina*). The tentacled, fringed hood of this nudibranch often causes it to be mistaken for some strange jellyfish caught up in eelgrass or kelp beds. Rather, the "bell" is a feeding apparatus designed to capture small organisms floating in the water. No gills are present, only a number of broad, spatula-shaped cerata down the back. The species grows to 10 centimetres (4 inches) in length. A careful look at eelgrass blades in shallow water in summertime will generally produce one of these.

through the mud covering 46 centimetres (18 inches) in eighteen seconds. For most others, progress is slow.

A cockle is also able to use its foot as a springboard to bounce itself across the sea bottom in search of a suitable site to dig itself in.

Scallops have developed another means of locomotion. They neither burrow nor crawl but swim by rapidly clapping their valves together, forcing water out of the mantle cavity in spurts of jet propulsion. The image created is one of a disembodied pair of dentures biting its way through the sea.

Mussels, so common in great numbers on wharf pilings and rocky shores, have modified the foot into a spinning device used to anchor the mussel shell by guy wires in much the same way as a spider spins a web. A byssus (filament-producing) gland secretes a mucus, which flows along a groove in the foot to the point where it is to be attached. The mucous thread hardens immediately on contact with sea water. A great number of threads are thus spun until the attachment is secure, often forming a great brush of stiff threads. The rock oyster or jingle shell *(Pododesmus cepio)* has a special pear-shaped window in its lower shell through which the byssal attachment is made.

Rarest of all fabrics, the "cloth of gold" of ancient times was spun from the byssus threads of a Mediterranean clam known as a pen shell *(Pinna nobilis)*. The material so produced is said to have had a silky sheen and was olive-gold in colour.

A structure often confused with the foot is the siphon, or the "neck," as it is known among clam diggers. Its function is that of a double-barrel snorkel, allowing the claim to sit comfortably buried in mud or sand while having access to the water above without taking in bits of sand, silt and gravel. One tube of the siphon brings water into the clam and is known as the incurrent siphon. The other tube takes water out of the clam and is known as the excurrent siphon. In most species with a siphon, the two tubes are joined. Occasionally, as in the bent-nosed clam *(Macoma nasuta)*, the siphons are separate hoses, the incurrent one being used to vacuum the mud surface in a large arc for bits of detritus that may have settled nearby. If the siphons were joined, one would be sucking in bits of food and the other one would be blowing the food away at the same time.

Siphon lengths offer a valuable clue to the depth where the particular species are found. For example, the littleneck clam *(Protothaca staminea)*, with its short siphon, occurs very near the surface. The geoduck *(Panopea generosa)* has a long siphon, allowing it to bury itself from 75 to 150 centimetres (30 inches to 5 feet) down. It stands to reason that because the scallops, oysters and mussels do not burrow, they do not require a snorkel arrangement.

By and large, bivalves and other marine animals, with the exception of marine mammals, breathe water as terrestrial animals breathe air. In most bivalves, respiratory gills feed the animal as well as absorb oxygen. A meal consists of tiny particles suspended in the water, such as diatoms, and minute plants and animals. Bits of food become trapped in the mucous coating on the gill filaments and are then moved along a groove by the beating of many tiny hairs known as cilia. Food passes into the mouth and along the digestive tract. This mode of eating is known as filter feeding. Not surprisingly, a great volume of water must be filtered to supply enough food, not just for ordinary sustenance of the animal but for growth as well. It has been calculated that a good-sized Japanese oyster *(Crassostrea gigas)* filters 13.25 litres (3½ gallons) of water per hour.

Life for most bivalves is simple, mainly because they do not possess the equipment needed to have it otherwise. Of the senses, they have organs of balance known as statocysts and will respond to touch by shutting their valves. There is no brain, only four pairs of ganglia, or nerve centres.

The scallop is able to perceive its environment visually. When its valves are agape, rows of tiny blue-green eyes peek from between the shells. Rows of feathery, sensitive tentacles fringe the shell margins. Its many eyes have both a retina and lens and are thought to register an image. Why the scallop is the only bivalve to be thus favoured may be explained by the fact that the free-swimming scallops are the most mobile and vulnerable of all bivalves and therefore need better sensory equipment than their cousins in order to avoid predation. It may be argued that the rock scallop *(Hinnites giganteus)* is an attached, sedentary species like the oyster and, therefore, should be exempt from special privileges. Yet if you look carefully at the upper shell of an attached

Littleneck clam (*Protothaca staminea*). Because the fused siphons of this clam are very short, individuals live just beneath the sand or gravel surface and are therefore easily harvested. The shells are moderately heavy and ribbed, and reach an average size of 6 centimetres (2¼ inches) in diameter. The species ranges from the Bering Sea to Mexico.

A small garden claw is the best tool for gathering this clam and the cockle. It disturbs much less sand and gravel than a shovel.

Cockle (*Clinocardium nuttallii*), also known as the heart cockle, inhabits quiet bays of mud or sand beaches and is buried just beneath the surface. It ranges from the Bering Sea to San Diego, from the high intertidal zone to a depth of 10 metres (33 feet). The cockle grows to 10 centimetres (4½ inches).

If you want to know what kinds of live clams occur in an area, try identifying the empty shells lying about. Chances are they are accurate indicators of what lies below the surface.

scallop, you will see the impression of a perfectly formed scallop shell, about 2.5 centimetres (1 inch) in diameter, whereas the rest of the large shell is often irregular or grossly misshapen. Until the scallop is the size of a silver dollar, it is a free-swimming, free-living creature. Then, for some obscure reason, it settles, attaching itself to a rock by its lower shell, and continues to grow, though not with the same symmetrical beauty that marked its youth.

Scallops, like all shell-bearing mollusks, enlarge their shells by adding shell material to the shell margins. Mollusks never shed old shells and regrow new ones, as do crabs, though they are able to patch holes or cracks in the shell if the damage is not extensive. Secretions from the mantle tissue below the shell add patching layers from within. Much the same mechanism is responsible for the production of pearls in mussels and oysters. Sand grains or parasite larvae are an irritation to the soft mantle tissue, just as a crack or hole would be. The animal responds by secreting layers of nacre (mother-of-pearl) over the irritation, the same porcelainlike finish of the shell's interior. With time and more layers, the tiny bead becomes a pearl, and in the pearl oysters (*Pinctuda*) is finished with an iridescent lustre of great beauty. A cultured pearl is derived from an oyster that has been deliberately "seeded" with a small grain to stimulate pearl production artificially.

Most bivalves are either male or female; some cockles are hermaphroditic, and some, such as the native oyster (*Ostrea lurida*) and the Japanese oyster (*Crassostrea gigas*), alternate from season to season or annually. Many spawn year round, others only when "inspired" by the warm waters of summer. Whatever method is followed, there is no doubt that many bivalves have a tremendous potential for reproducing themselves. For example, a single female Japanese oyster releases 500 million eggs during a season. No mating occurs; the eggs and sperm are simply shed into the water, and by some mysterious natural predisposition many of them meet, the eggs are made fertile, and they begin a free-swimming infancy, which may last hours or weeks.

Soon, however, life begins in earnest for those larvae that have escaped any number of dangers and predation. By the time a Japanese oyster is ready to settle at two to four weeks, it is fully formed at only 3 millimetres

(⅛ inch) in length. The settling of oysters is known as spatting—hence the name "spat" for the young oyster. An oyster will live in relative peace from two to five years before reaching an edible size of 11.25 centimetres (4½ inches), when it will most likely be harvested. If left alone, it may live to twenty years. It is estimated that most other bivalves have a natural life span of five to twelve years.

Bivalves are capable of producing a tremendous number of eggs, but few grow to maturity. The available food supply and space would be severely taxed if every year each female oyster successfully reproduced 500 million or even a thousand more oysters. Predation is very heavy not only at the juvenile stage but at all life stages of bivalves. With the exception of the scallop, which may be able to see its enemy coming and flee from capture, or the cockle, which may bounce its way out of a starfish's reach, the bivalves have little ability to anticipate and avoid capture. For many, their only defence is to remain buried, close their shells and hope for the best.

Enemies are everywhere. Shore birds, fish, seals, mink, otter, sponges, starfish, octopus and people are but a few. And, of course, the drilling snails are also hostile; not only the huge, bulbous moon snail (*Polinices lewisii*) but the diminutive eastern drill (*Urosalpinx cinerea*) and the Japanese oyster drill (*Ocenebra japonica*), both accidentally introduced with the Eastern (*Crassostrea virginica*) and Japanese oysters (*Crassostrea gigas*).

West Coast bivalves are common to many habitats— the shifting sand of surf-pounded beaches, quiet estuaries and protected bays, exposed rocky shores, sand, gravel and mud. A few species even bore into rocks. The teredo, discussed later in the chapter, lives only in wood.

Horse clam or gaper clam (*Tresus capax*) is the commonest species to be found on tide flats of quiet bays. It may grow to 20 centimetres (8 inches) and has a large siphon used to transport water from above the muddy substrate to the clam, which may be up to 50 centimetres (20 inches) below. The horse clam occurs intertidally and below from Alaska to California. This clam is often confused with the larger and less frequently found geoduck (*Panope generosa*).

Clams are harvested by human clam diggers, moon snails and some sea stars. The commensal pea crab is often found in the mantle of the horse clam.

SOME NOTES FOR SHELLFISH HARVESTERS

Theoretically, all clams, mussels and oysters are edible. However, you must pay attention to the legal size limits applicable to some species in certain areas and to possible contamination by pollution or red tide.

Shellfish found adjacent to sewage outlets or industrial effluents should not be harvested. Such areas often have posted warnings. If an area is at all suspect, avoid it.

Red tide is mentioned earlier. It is a temporary summer phenomenon, though it may infrequently occur in winter, rendering many filter-feeding organisms, such as shellfish, toxic to human beings. The planktonic organism responsible *(genera Gonyaulax)* proliferates to such a degree that the water becomes coloured by its density. The filter feeders strain the organisms from the water in the normal course of their feeding and tend to concentrate in their flesh a toxic material from the planktonic organism in their flesh. If eaten, these shellfish can cause extreme intestinal discomfort. In particular, butter clams and California mussels are to be avoided during a red tide and even up to a year afterward.

Shellfish harvesters are advised to contact local Department of Fish and Wildlife officials regarding legal size, bag limits and areas closed to harvesting.

Some species of shellfish are less tasty during certain times of the year as a result of their sexual cycle. This does not mean they are poisonous, only that the flesh is not in prime condition. Oyster flesh becomes soft, thin and watery during spawning and for a period afterward in the summertime, quite unlike the firm flesh during the colder months. Clams are at their table best when in a prespawning condition during the winter time. In the summertime, after spawning, they are tough and less tasty.

One final word of caution to the shellfish harvester: take only what you need, and when digging in the sand or mud for clams, always fill in the holes after the shellfish have been gathered. Many other animals living in the sand, mud and gravel depend on your doing so. If disturbed sand or mud is not replaced, many shellfish larvae and other animals may be washed free of their protection in the sand and die at the mercy of the surf.

Mussels. Mussels are characterized by a thin, somewhat kidney-shaped slate-blue, black or dark brown shell. They attach themselves to pilings, rocks or each other by many threads known as the byssus. They are best steamed in the shell or in fish soup, shells and all.

Scallops. Scallops, or pectens, are recognized by their "Shell Oil" shell—a round fan with two "ears" at its base. Generally, only the large adductor muscle is eaten after being sautéed whole or thinly sliced, gently pounded and then fried.

Sand clam (*Macoma secta*). Long separate siphons allow the sand clam to live well below the sand's surface, in depths from 20 to 40 centimetres (8 to 16 inches). This species appears to prefer clean sand. The incurrent siphon is very mobile, sweeping the surface of the sand above the clam, like a vacuum cleaner, picking up bits of food. Sand clams grow to 10 centimetres (4 inches) and live in sand from the intertidal zone to depths of 30 metres (98 feet) from southern Alaska to Mexico. Because it lives so deeply buried, this clam seems to have few predators.

Razor clam (*Siliqua patula*) is olive-green to brown and grows to a length of 15 centimetres (6 inches). It occurs on surf-swept beaches of the open coast from the Arctic Ocean to California in the low intertidal zone to several fathoms. The razor clam has a tremendous capacity to dig quickly into the sand below if it is disturbed in its resting place just below the surface. This clam is eaten by flatfish and the starry flounder and is a great favourite of human clam eaters.

Edible or blue mussel (*Mytilus edulis*) grows to 10 centimetres (4 inches) and is found in masses on rocks, pilings or some other firm attachment. It occurs intertidally from the Arctic to California. Colour may vary from black-blue to brown. This species does well in quiet waters of low salinity. Purple sea stars, predaceous snails, crabs and birds all eat blue mussels. In addition, thick beds of this animal provide hiding places for all kinds of other smaller intertidal animals.

California mussel (*Mytilus californianus*) is a large black mussel of 15 to 20 centimetres (6 to 8 inches) or larger and is found attached to rocks on the surf-swept open coast from Alaska to Mexico. The flesh is bright orange and edible. The California mussel frequently contains pearls; however, these are of no commercial value. A commensal pea crab is often found living in the mantle cavity of this species. Mussels are particularly affected by red tide.

Pacific or Japanese oyster (*Crassostrea gigas*) was originally introduced to the West Coast from Japan in 1902 or earlier. It is now found intertidally from northern British Columbia to California over many kinds of beaches. It may grow to 30 centimetres (12 inches) and is the basis of the British Columbia oyster industry. A grey to white heavily fluted shell distinguishes this species from other oysters. Recreational harvest may be regulated, so check for restrictions or daily bag limits.

Rock oyster or jungle shell (*Pododesmus cepio*). The shells of this bivalve are nearly circular. A large hole in the centre of the lower valve accommodates a heavy byssus, which anchors the animal firmly to a substrate. The rock oyster grows to 10 centimetres (4 inches) across and occurs from the low tide level to a depth of 30 fathoms. It ranges from the southern Bering Sea to Mexico. It is often camouflaged by liberal growths of plants and animals on its exposed valve. The flesh is bright orange and considered good eating.

Rock scallop (*Hinnites giganteus*). The juvenile of this species is free swimming for a time before settling and becoming attached by its lower valve. The upper valve is heavy, coarsely ribbed and often overgrown with encrusting organisms. The rock scallop ranges from Alaska to California in the low intertidal to a depth of 25 fathoms. It grows to a length of 25 centimetres (10 inches). Growth is slow, and it may take twenty-five years for a rock scallop to reach maximum size.

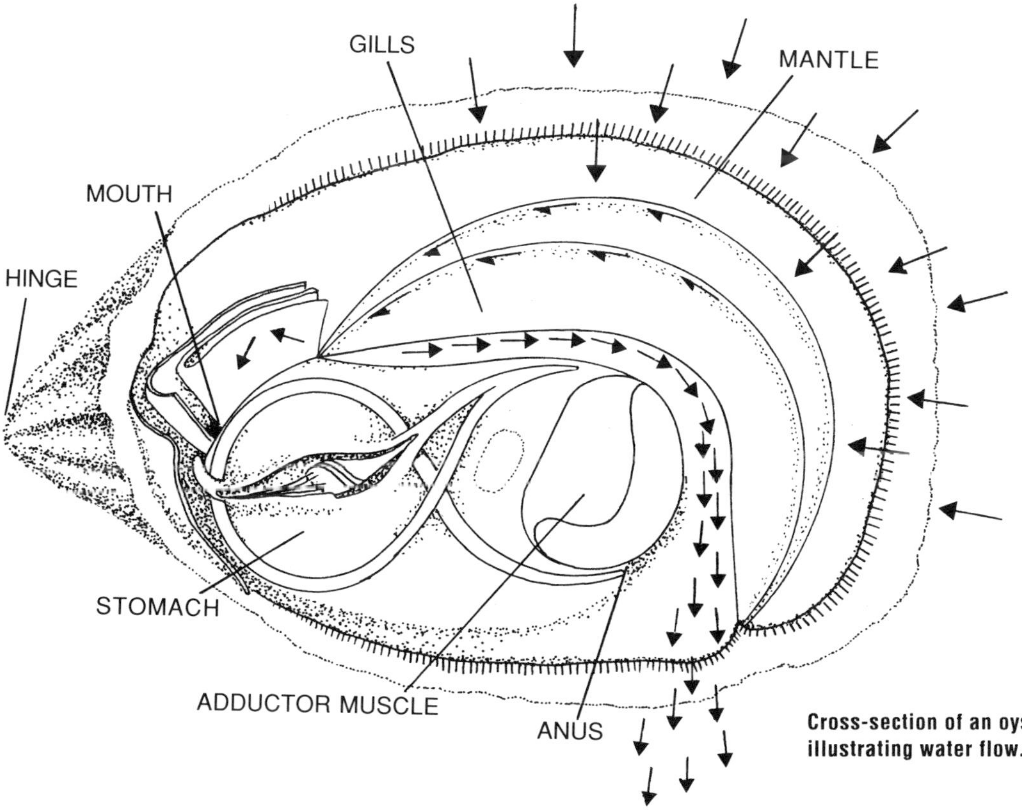

GILLS

MANTLE

MOUTH

HINGE

STOMACH

ADDUCTOR MUSCLE

ANUS

Cross-section of an oyster, illustrating water flow.

Oysters. Oysters generally have a white or grey-white frilly shell. They cement themselves directly to a substrate and do not have byssal threads. They are delicious eating raw, stewed or fried.

Clams. Clams and cockles do not generally attach themselves to a substrate but bury themselves in the sand, mud or gravel. They do not have byssal threads, nor do they swim, but they move using a muscular foot. They are eaten fried, stewed or in chowders. All species are edible.

THE WAMPUM OR MONEY TUSK SHELL (SCAPHOPODA)

A particularly conscientious beachwalker may be fortunate enough to find a money tusk or wampum shell *(Dentalium pretiosum)* washed up on the beach of some West Coast shore. Certainly this animal will not be found alive intertidally, as it is a subtidal to deep-water species, preferring depths of 9.15 to 76.25 metres (30 to 250 feet) along its range from British Columbia to San Diego.

Estimates of tusk shell species throughout the world range from 350 to 1000. Several species other than the above-mentioned are recorded on the West Coast—for example, *Dentalium dalli* and *D. rectius.*

Unlike the elephant tusk, for which this group is named, the tusk shell is open at both ends. Burying itself horizontally in soft mud or sand with its narrow end just breaking the surface, the animal draws water in and out through the exposed opening and thus respires. Tusk shells have no gills and absorb oxygen through tissue lining in the mantle cavity.

Emanting from the submerged, larger opening are dozens of prehensile threads or tentacles, which search the sand for tiny single-celled animals. Once caught in the threads, food organisms are drawn into the mouth and digested. Like burrowing bivalves such as clams, tusk shells have a muscular foot.

These obscure mollusks are of interest not only biologically but also historically. During the early days of European contact with native people of the West Coast, the money tusk, or wampum, was used as one form of money, and according to trading records of the time, a 5-centimetre (2-inch) shell had a purchasing power equal to a shilling.

Wampum shells had been in use as currency by the native people long before the arrival of Europeans. Value was determined by length. A 2.5-centimetre (1-inch) shell had little value; a 5-centimetre (2-inch) shell had greater purchasing power; and a 7.5-centimetre (3-inch) shell was of such worth that it was owned only by wealthy chiefs.

Wampum shells, also known as dentalia shells, were gathered by canoe from known beds using a tool designed to plunge into the mud and trap the shellfish. Naturally, the locations of such beds were closely guarded secrets.

Wampum shells can be seen in museums on some of the ceremonial costumes of the coastal tribes. In keeping with the prestige and value of the shells, they were often incorporated into the design of dress worn for special feasts and dances.

CHITONS OR SEA CRADLES (AMPHINEURA)

The still, low profile of the chitons means they are often overlooked among the rich and colourful abundance of intertidal fauna. This is despite the fact that at least seventy-four of one thousand known species occur on the West Coast from central California to Alaska. All chitons are marine, and most occur intertidally or in shallow water on rocky shores.

It is not surprising that the chiton is overlooked. It often bears many small plants and animals on its back and attaches itself so firmly to its grounding that it fairly melts into the environment.

The chiton takes the shape of a low, oval mound and bears on its back six to nine, but generally eight, flat shells that overlap each other in series from front to rear like broad shingles. The shells are secreted by the mantle tissue beneath and present a solid shield to most of the creature's exposed surface. The shells are held firmly in place by a girdle surrounding the whole, much like an elasticized belt. The girdle may be smooth, spiny, hairy or, as in the case of the giant gumboot chiton (*Cryptochiton stelleri*), enlarged to such an extent that the shells are totally concealed by the thick, gritty girdle. Chiton shells are not shed to allow for growth. As in all shelled mollusks, the mantle enlarges the shells at their edges as the animal grows.

The bottom of the chiton is a broad, muscular foot.

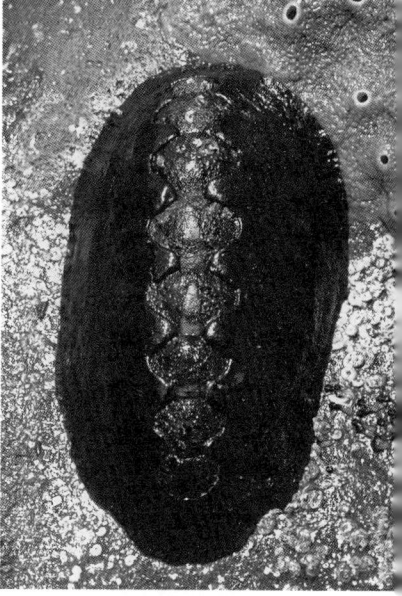

Leather chiton (*Katharina tunicata*). A smooth, leathery black girdle almost covers the eight shell plates of this common intertidal chiton. The species is a herbivorous grazer of rocky beaches from California to Alaska. It grows to 12.5 centimetres (5 inches) and has an estimated life span of three years. Look for this chiton on the top or sides of rocks exposed to strong wave action.

Gumboot chiton (*Cryptochiton stelleri*) is the largest chiton in the world, growing to 30 centimetres (12 inches). All shell plates are covered by a thick, gritty brick-red girdle. The gumboot chiton is fairly common yet difficult to see in the low intertidal zone of rocky coasts. It ranges from Japan to the Aleutian Islands and down the coast of North America to California. Gumboot chiton eats algae, grows slowly and probably lives for twenty years or more. It has few natural enemies.

Using a combination of a sticky secretion and vacuum-like suction, the chiton clamps down securely on its rocky substrate. Because the chiton cannot withdraw into a protective shell as can many other mollusks, it must rely on the tenacity of its grip to foil would-be predators. If removed from their foothold, chitons will tend to curl inward, forming a kind of cradle—hence the term "sea cradle" used to describe the chiton in some localities. Besides providing attachment, the chiton foot facilitates slow movements by backward waves that push the animal forward at a rate of 1.25 to 15 centimetres (½ to 6 inches) per minute. Observers will seldom see chitons moving during the day, since they are nocturnal creatures that venture abroad in search of food when the light is dim and return to their "homing" area at daybreak.

Chitons are vegetarians. At the head end of the animal is a poorly defined head, beneath which is the mouth and a special feeding tool, the radula. The radula is a straplike tongue bearing many tiny teeth and is used to rasp the algal film from rocks. When not in use, the radula is tucked away into a pouch leading off from the digestive tract.

The chiton lacks eyes or tentacles on its head, yet it is able to sense touch and perceive light variations via sensory cells held in pits on the shell plates. In this way the chiton "knows" when darkness has fallen and it is safe to search for food or when it must clamp down firmly to avoid being eaten. Other sensory cells in the form of chemical receptors in the mouth give the chiton a sense of taste.

Respiration is accomplished by gills, located in two lateral grooves between the chiton's foot and its shells. Water entering these grooves at the head end passes over six to eighty pairs of gills (depending on the species), where oxygen exchange takes place, and exits to the rear of the animal. Absorbed oxygen from the water and nutrients from digested food are distributed throughout the body internally by a simple circulatory system.

Too little is known of the life histories of individual chitons to allow for broad, generalized statements about life span or reproductive cycle. It is known, however, that sexes are separate and fertilization of eggs takes place in the mantle cavity of the female. Chitons tend to be gregarious during periods of breeding, and therefore the likelihood that a female will draw in sperm with the water normally taken in for respiration is greatly enhanced. As the water and sperm pass through and out of the mantle cavity, the eggs are fertilized. Generally the eggs are then laid either singly, in strings or in a jelly mass.

The gumboot chiton, for example, breeds in the spring, laying its eggs in two long spirals of jelly. After hatching, the free-swimming larvae begin to settle in a matter of hours and proceed to metamorphose into shell-bearing adults.

The gumboot chiton is the largest of all chitons, growing to over 30 centimetres (1 foot) in length and half as wide. It is common intertidally throughout its range from the Bering Sea to California, yet because of its covered shells and large, bulbous size, it is often not recognized.

A number of other, smaller chitons bearing the distinctive eight shell plates are very common intertidally as well. Among these are the lined chiton (*Tonicella lineata*), brightly coloured with dark brown lines zigzagging over a lighter background; the mossy chiton (*Mopalia spp.*), characterized by a bristly girdle, and "black Katy," or the leather chiton (*Katharina tunicata*), distinguished by a black, leathery girdle covering the edges of the shells.

Hairy chiton *(Mopalia lignosa)*. This chiton is fairly common on the sides or under boulders at low tide. It is typically found out of direct sunlight. Small, light-sensitive organs occur on the shell plates, and laboratory tests show that the chiton retreats from light. A small animal, to 7 centimetres (2¾ inches), it ranges from Alaska to California in bays and on open coasts.

TEREDOS (SHIPWORMS)

Humans view such mollusks as the butter clam, oyster and abalone with benign indulgence and a stirring of gastronomic anticipation. How delicious are these succulent edibles of the sea!

It seems only fair that a group of animals such as these, which have been heavily harvested by human beings since well before the time of recorded history, should include a creature capable of creating great destruction and misery for humankind.

The shipworm, or teredo, enemy of seagoing people through the centuries, is the mollusk's revenge. The shipworm is not a worm at all but a very unclamlike clam. "Teredo" is the name applied to some sixty-six different kinds of wood-boring clams, two of which are active on the West Coast. These are the giant *Bankia setacea* and the smaller *Teredo navalis*. These mollusks relish submerged wooden structures, be they wharves, boat hulls or logs. They bore and tunnel their way through all kinds of wood with frightening speed, rendering their wooden host a crumbling mass in a matter of months. The North American Pacific coast endured a famous case of teredo infestation from 1917 to 1920, when wharves and jetties in San Francisco Bay suffered losses in excess of $25 million in 1917 dollars.

Like other clams, teredos begin life as free-swimming larvae with a pair of shells. For the tiny teredo, life will cease within weeks if it does not come to rest on submerged wood. If a wooden surface is found, the teredo, only 0.005 millimetre (⅕₀₀₀ inch) in diameter, searches for an appropriate site to begin its boring. Using the edges of its fragile shells, the teredo pivots its paired shells to rasp a minute puncture in the wood and burrows beneath the surface.

Now the real work begins as the shells harden, developing filelike rasping edges (the better to dig a bigger burrow) to accommodate the rapidly growing teredo. With little external sign of its inner presence, the teredo begins burrowing in earnest. By three months of age, the teredo is chewing its way at a rate of 18 millimetres (¼ inch) per day. Generally, the burrow follows the wood grain but will shift in direction to avoid knots and bolts in the wood or a neighbour's burrow.

As the teredo grows, it extends backward out of its shell so that the rasping shells remain like jaws at the creature's head end, with the body trailing behind in a wormlike fashion—hence the name "shipworm."

Like other clams, the teredo must provide for water circulation over the gills. To this end, a pair of long siphons, or tubes, extends from the tail end of the teredo to the opening of the burrow. Water entering through the incurrent siphon brings with it oxygen and planktonic particles for food, which are absorbed as the water passes over the gills. Water exhaled through the excurrent siphon carries away with it metabolic wastes and, during the breeding season, eggs, sperm or larvae.

What happens to the wood that the teredo displaces as it burrows? The "sawdust" is passed through the digestive tract of the teredo and by the action of two enzymes in the stomach is, in part, converted into food for the shipworm. The teredo balances its diet of woodchip soup with additional planktonic material strained

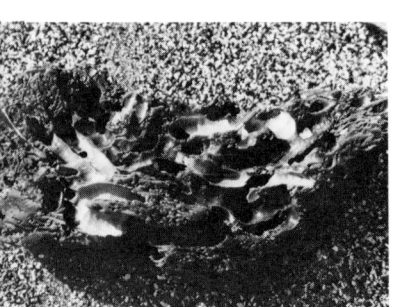

Wood showing the burrows of the large shipworm *Bankia setacea*. Shipworms and pile worms are the most highly modified of the boring clams. Once young animals have penetrated a wood surface—wharf piling, ship's hull or boomed log—the clam bores, following the grain of the wood. If not crowded, this species can grow at 7.5 centimetres (3 inches) per month. If heavily infested, most untreated wood is destroyed within a year.

from the water in the conventional clam manner.

As it bores, the teredo's mantle secretes a limy lining on the interior of the burrow. It also secretes a pair of calcareous paddles known as pallets, which are situated at the tail end of the animal, near the entrance of the burrow. When the siphons are retracted, the pallets are held against the burrow's opening, thereby securely sealing the teredo in its home during unfavourable periods of water contamination or exposure. This protective device makes it extremely difficult to eradicate teredo infestation through the use of external treatments. Shipworms are very sensitive to unusual chemicals and respond to them by retreating into protective seclusion.

Once a shipworm has penetrated the wood's surface, it will not burrow to the outside, nor can it be induced to leave its burrow to avoid noxious chemicals. Thus far the only chemical able to penetrate the teredo's defence is a solution of an arsenic compound. Otherwise, the only means of killing shipworms is to remove the affected wood from the ocean. Without access to water and the oxygen it carries, the animal dies of suffocation.

The control of shipworms is extremely important on the West Coast, particularly in areas where coastal waters are used to store and transport floating wood. Both *Bankia setacea* and *Teredo navalis* have caused considerable loss and damage to log booms in British Columbia, Washington and Oregon.

Bankia is found from Kodiak, Alaska, to San Diego and is responsible for the larger bore holes frequently seen in beach drift logs. This species grows to 90 centimetres (3 feet) in length and can be as thick as a person's finger. *Teredo navalis* is half the size of *Bankia* and can be recognized by its smaller borings. *Teredo navalis* was thought to have been introduced to the western North American coast from infected ships arriving from the Atlantic coast. Now it appears to be everywhere.

Both species of teredo are fast-growing and prolific. *Bankia* reproduces throughout the late fall and winter, shedding eggs and sperm directly into the water through the excurrent siphon. Reproductive cells meet in the water and produce a fertilized larvae, which, in turn, begins boring within weeks. Interestingly, the *Bankia* shipworm matures sexually first as a male, later developing into a sexually reproductive female.

Teredo navalis is not so casual in its reproductive efforts. The female of this species broods her fertilized larvae in her mantle cavity until the young are a few weeks old. Brooding occurs in the late summer and fall, ensuring that all offspring will be well-housed by Christmas. Like the *Bankia* shipworm, *Teredo navalis* breeds continuously for a number of weeks, the female producing nearly a million eggs.

Since no species of wood is teredo-resistant in all geographical areas, protection from this insidious pest continues to be a problem. Many submerged structures that have traditionally been made of wood are now made of steel and concrete. Metal sheathing on wood pilings and boat hulls is also used to protect wood, and other submerged wood is coated with creosote to discourage teredos from settling. Because teredo infestation cannot be fully determined on casual examination, divers may use sonic equipment to test the practical strength of submerged structures. The forest industry uses testing stations where submerged wood blocks and known breeding habits make it possible to forecast infestations. This allows loggers to move logs to fresh water, treat logs with chemical sprays or otherwise pull logs before damage occurs.

OCTOPUSES AND SQUID (CEPHALOPODS)

What has two eyes, ten arms, is 17 metres (56 feet) long and weighs many tons? The answer is *Architeuthis* of the North Atlantic, better known as the giant squid. (Perhaps it exists in the Pacific, too, since remnants of the species have been found in stomachs of sperm whales taken off the California coast.) Horrific and gigantic, the inspiration for medieval tales of the sea monster "Kraken," this creature does in fact exist and is the largest of all living invertebrates.

The fossil record shows over ten thousand cephalopod forms, yet only seven hundred species exist today, all of which are marine. This decrease may indicate that cephalopods are not destined to become a dominant living form. Although they are not a prominent animal group, their life history, behaviour and appearance are unique in many ways. Thus, the cephalopods deserve more than cursory mention. Discussion here will be lim-

ited to the octopus and squid. The shelled nautilids will not be considered, since none occur on the West Coast.

Octopuses and squid deviate in very many ways from the generalized mollusk form, as typified in snails and clams. Most obvious are the apparent loss of the protective shell so tenaciously retained by the majority of mollusks and the development of many suckered arms. The squid has discarded all but a simple internal stiffening rod known as the pen; the cuttlefish, close cousin of the squid, has retained just enough shell to provide bird cages with cuttlebone; and the octopus bears no vestige of shell at all.

However, the loss of shell can by no means fully explain the peculiar physique of the octopus and squid. To the uninitiated, these animals appear to be all head, with a multitude of suckered arms where a neck and body should be. What appears to be head is a functional baglike body containing all the organs and equipment needed to survive in a competitive environment. Thus, another reason that these animals look strange is that their bodies are enclosed in a mantle. The mantle covers structures such as gills and alimentary openings

Stranded squid found on a beach in the Queen Charlotte Islands, British Columbia, probably the North Pacific giant squid (*Moroteuthis robusta*). This animal is occasionally seen at the surface near the surf line or on shore. It is typically an offshore animal living at depths of 100 to 600 metres (328 to 1970 feet) from Japan to southern California. The body grows to 1.2 metres (4 feet) in length, with another 3.3 metres (11 feet) of tentacles. It is a very lucky beachwalker indeed that comes across the large carcass of some offshore animal and has the opportunity to look at the specimen before the tide and any army of underwater scavengers recycle the remains.

by enclosing them in a mantle cavity and giving the body a smooth, globular shape.

Quite simply, the mantle is the skin on the outside of the body, excluding the arms. It can be likened to a reasonably but not quite perfectly fitting bag that has been pulled over the viscera or organs to the point where the arms are attached. This bag is attached to most of the viscera, leaving an open, ample pocket to one side. The pocket, or pouch, is the mantle cavity. The opening to the mantle cavity of a living octopus or squid can be seen as a flap just short of the arms, or in front of them.

Squid are adept swimmers, octopuses less so. Propulsion in these animals is typically provided not by the arms but by the mantle cavity. Water is drawn into the cavity, and the point of its entry is then closed off by a one-way, valvelike arrangement. Strong muscles surrounding the mantle cavity contract, forcing the water out through a funnel. This action causes the animal to shoot forward in spurts, in a kind of jet propulsion. The outflow funnel is flexible, and its orientation can be altered, allowing the animal to change direction and to swim backward or forward.

Squid spend most of their time swimming and as a result have developed lateral stabilizing fins. Octopuses are more prone to be bottom dwellers, and they crawl over the ocean floor on their arms with remarkable facility, as well as taking to the open water with swimming motions.

Octopuses and squid are carnivorous. They stalk live prey and, having somewhat different food preferences, exhibit different methods of capturing their food.

Squid capture fish and shrimp "on the wing." To do this, they use two long tentacles. The tentacles are much longer than their other eight arms and are capable of great elastic extension. In some species the tentacles are armed with clawlike hooks in addition to sucker disks elevated on short stalks. To capture its prey, a squid shoots out its tentacles, which draw the prey into the firm grasp of eight waiting arms. The regular arms are also well supplied with sucking disks circled in horny rings.

Octopuses do not have the squid's two extra tentacles for capturing prey. Instead, this animal stalks its quarry, generally during the evening hours or at night, and at the right moment descends like an umbrella on its vic-

Giant Pacific octopus (*Octopus dofleini*). Handle with care because even though octopuses have no bones, scales or other hard parts, this animal has a powerful, parrotlike beak in its mouth, where its eight arms meet on the underside. Giant Pacific octopus are not aggressive but, like nearly all animals, will attempt to bite when handled.

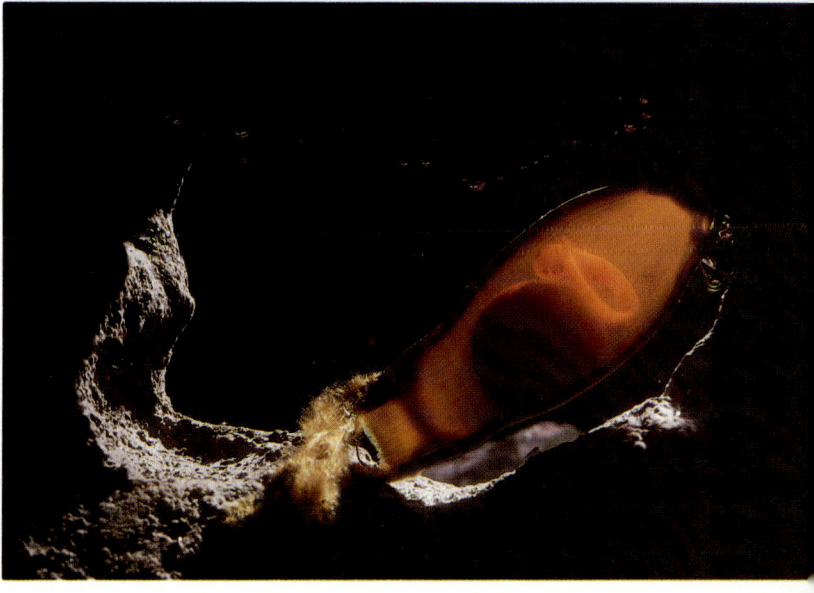

Plate 1
Harbour seal
(*Phoca vitulina*).

Plate 2
Brown cat shark egg case
(*Apristurus brunneus*).

Plate 3   Brittle stars, class Ophiuroidea.

Plate 4
Crimson anemone
(*Cribrinopsis frenaldi*).

Plate 5
Purple or ochre star
(*Pisaster ochraceus*), Helby
Island, British Columbia.

Plate 6
Sea stars (*Pisaster ochraceus*)
and green surf anemones
(*Anthopleura
xanthogrammica*),
Ruby Beach, Washington.

Plate 7   Juvenile sunflower star (*Pycnopodia helianthoides*).

Plate 8   Lined chiton (*Tonicella lineata*).

Plate 9   Steller's sea lions (*Eumetopias jubata*), Hazy Island, southeast Alaska.

Plate 10   Fluffy sculpin (*Oligocottus snyderi*).

Plate 11   Sea pen (*Ptilosarcus guerneyi*); detail of feeding polyp.

Plate 12
Clam siphons, sea squirts
and bryozoans.

Plate 13
Head of pipefish
(*Syngnathus griseolineatus*).

Plate 14
Grunt sculpin
(*Rhamphocottus richardsoni*).

Plate 15
Longfin sculpin
(*Jordania zonope*).

Plate 16
Juvenile red snapper
(*Sebastes ruberrimus*) and the
strawberry anemone
(*Corynactis californica*).

Plate 17
Coon-striped shrimp
(*Pandalus danae*).

tim. Octopus sucker disks are not stalked, nor do they possess horny rings. They are capable of enormous suction, and once caught, prey is unlikely to escape.

Both squid and octopuses have a powerful, horny beak at their mouth, something like a parrot's beak. As the prey animal is held in the squid's ten arms or the octopus's eight arms, the beak bites the prey. Glands containing poison saliva discharge their fatal juices into the wound made by the beak, ensuring a swift dispatch of the luckless victim. Food thus captured cannot be chewed with teeth and jaws, as there are none, but a modified radula serves the same purpose. In most mollusks the radula is a toothed ribbon that extends out of the animal's mouth and is used to scrape algae off rocks. In the octopus and squid, the radula has broadened to grind flesh before it is passed to the digestive tract for rapid digestion. Waste materials from the digestive process exit the body through the anus, which opens into the mantle cavity. Wastes are flushed away by water brought in to irrigate the gills, which hang in the same cavity.

As squid and octopuses hunt, so they, in turn, are hunted. Squid are eaten by all manner of fish and marine mammals, as are octopuses, particularly when they are juveniles. The giant squid is preyed upon by the huge sperm whales. Captured sperm whales frequently bear large circular sores inflicted by the sucker disks during battles with the giant cephalopod.

Both squid and octopuses have evolved some means of defence that are unique to their species. One mode of defence is the ink gland. When intimidated, they will eject an inky secretion from a gland located behind the anus. This creates a dense cloud in the water, which acts as a smoke screen to allow the animal to escape while also paralyzing the olfactory sense of a pursuing fish. Since whales have no sense of smell, it is unlikely that the ink cloud would do more than visually mask a cephalopod from that predator.

Many octopuses have yet another protective mechanism: the ability to change colour, using pigmented cells known as chromatophores. By expanding or contracting different pigment cells, of brown, black, red, yellow and red-orange, the octopus can perfectly reproduce the colours surrounding it, making itself invisible when immobile.

Newly hatched Pacific octopus (*Octopus dofleini*). This individual hatched after sixty days in 12.8° C (55° F) water. At hatching, the length of the mantle measured 3.29 millimetres (³⁄₂₀ inch).

The well-developed eye and brain of the octopus and squid are helpful in hunting and in eluding enemies. These eyes are a long way from the simple light-sensitive organs of most mollusks and are remarkably similar to vertebrate eyes. The eye registers an image and, in conjunction with a fairly sophisticated brain, is able to distinguish between shapes and retain these distinctions as memories. In other words, squid and octopuses can learn and remember. Octopuses and squid are also the only mollusks known to sleep regularly.

Octopuses and squid do not hear, but they possess sensory capabilities other than vision. Touch is highly developed in the suckers, which receive tactile sensations. Smell appears to be registered in small pits beneath the eyes. All senses considered, the squid and octopuses are better equipped than any of their molluskan relatives to perceive the world in which they live.

Some years ago, octopuses were thought to be infested with a long parasitic worm whose back end trailed repulsively out of the mantle cavity. It has since been discovered that the "parasitic worm" occurs only in females as the aftermath of copulation and is the end of a male octopus's arm. Losing an arm, or rather part of one, is not as brutal as one would imagine because octopuses have the power to regenerate lost parts, as do many invertebrates. The male's arm is known as the hectocotyle arm and is adapted specifically for reproduction.

Octopuses and squid do not follow the impersonal molluskan habit of releasing eggs and sperm haphazardly to the vagaries of tide and wind, without so much as an acknowledgement passing between potential parents. The males personally present the goods to the female by hand, so to speak. Sperm is delivered neatly packaged in a long envelope known as a spermatophore. The arm adapted for this presentation has no suckers on its end. Its sole function is to take the spermatophore from its owner's mantle cavity and place it in the mantle cavity of the female, where the sperm are released to fertilize eggs.

Female squid do not brood their eggs. Egg clusters are generally secured at various points on the sea floor and left, completing the female role in the reproductive process. The eggs of these clusters are embedded in a gelatinous substance that is thick enough to protect the developing embryos from fungus attack and disagreeable enough in taste and odour to repel would-be predators.

Octopus mothers are more solicitous than squid mothers. They attach their eggs in strings to the top and sides of their rocky dens, staying with the eggs and caring for them until they hatch.

In a well-documented case of a giant Pacific octopus (*Octopus dofleini*) that spawned in captivity, it was noted that spawning occurred forty-two days following mating. Thousands of rice-size eggs were laid, covering an area of 0.19 square metre (2 square feet). During the weeks that followed, the mother octopus manipulated the egg mass with the tips of her tentacles, presumably cleaning the eggs to prevent fouling. In contradiction to other reports, this mother continued to eat during the brooding period, though she used her funnel to blow away any debris that happened to drift near the egg mass.

The brooding period—the time between egg laying and hatching—appears to vary according to the water temperature. The eggs of the giant Pacific octopus are generally brooded for five to six months. The mother usually dies after they hatch.

In octopuses and squid the egg hatches directly into a miniature adult without passing through a free-living larval stage. Depending on the species, they have an expected life span of somewhere between two and five years.

The giant Pacific octopus is the largest of all living octopuses and may grow to a weight of 45 kilograms (99 pounds), with a spread of 5 metres (16 feet) from arm tip to arm tip. An unusually large specimen was reported to have an arm spread of 9 metres (30 feet).

The giant Pacific octopus is common from the low intertidal to depths in excess of 24 metres (79 feet) in southern British Columbia and northern Washington. This species is known to occur as far north as the Bering Sea and very likely is found as far south as northern California. It is seldom seen in the southern areas, however, possibly because the animals prefer cold water and would have to go to very deep water in those areas. For the same reason, octopuses of southern British Columbia and northern Washington move into deeper water in the summertime.

Stubby squid (*Rossia pacifica*) is small. It grows to only 7.5 to 12.5 centimetres (3 to 5 inches). It ranges from Alaska to California and is occasionally found swimming at shore at night. Stubby squid crawl on their arms or swim. They dig in shallow bowls or a sandy sea floor, where they rest with their arms over their heads. They are found in the same habitat as their primary food, shrimp.

Adult Pacific octopuses live in rocky dens or crevices, and juvenile animals find security in discarded cans and jars or under rocks. The den is identified by an accumulation of crab shells and other debris at the entrance, indicating that the octopus kills its prey at the site of capture but does not consume it until after returning to the security and privacy of its own home. Though impressive in size, the giant Pacific octopus need not be feared in the wild. It is, by nature, a retiring, nonaggressive creature.

In addition to *Octopus dofleini*, at least two other octopus species are known from the West Coast. They are both small, the body of the larger being about the size of a large orange.

If you are lucky enough to find an octopus in the low intertidal, handle it with special care because it is a wonderful, shy and nonaggressive animal. Octopuses are unique in many ways, and to molest one is unthinkable.

Two other cephalopods are regularly encountered. These are the stubby squid *(Rossia pacifica)* and the opalescent squid *(Loligo opalescens)*. The latter is common in open waters of the West Coast of North America and is seldom seen except when the females come inshore to spawn. This species is generally between 22 and 50 centimetres (between 9 and 20 inches) in length and is harvested commercially for food and bait.

In total, twenty-four species of cephalopods have been recorded from central California to Alaska.

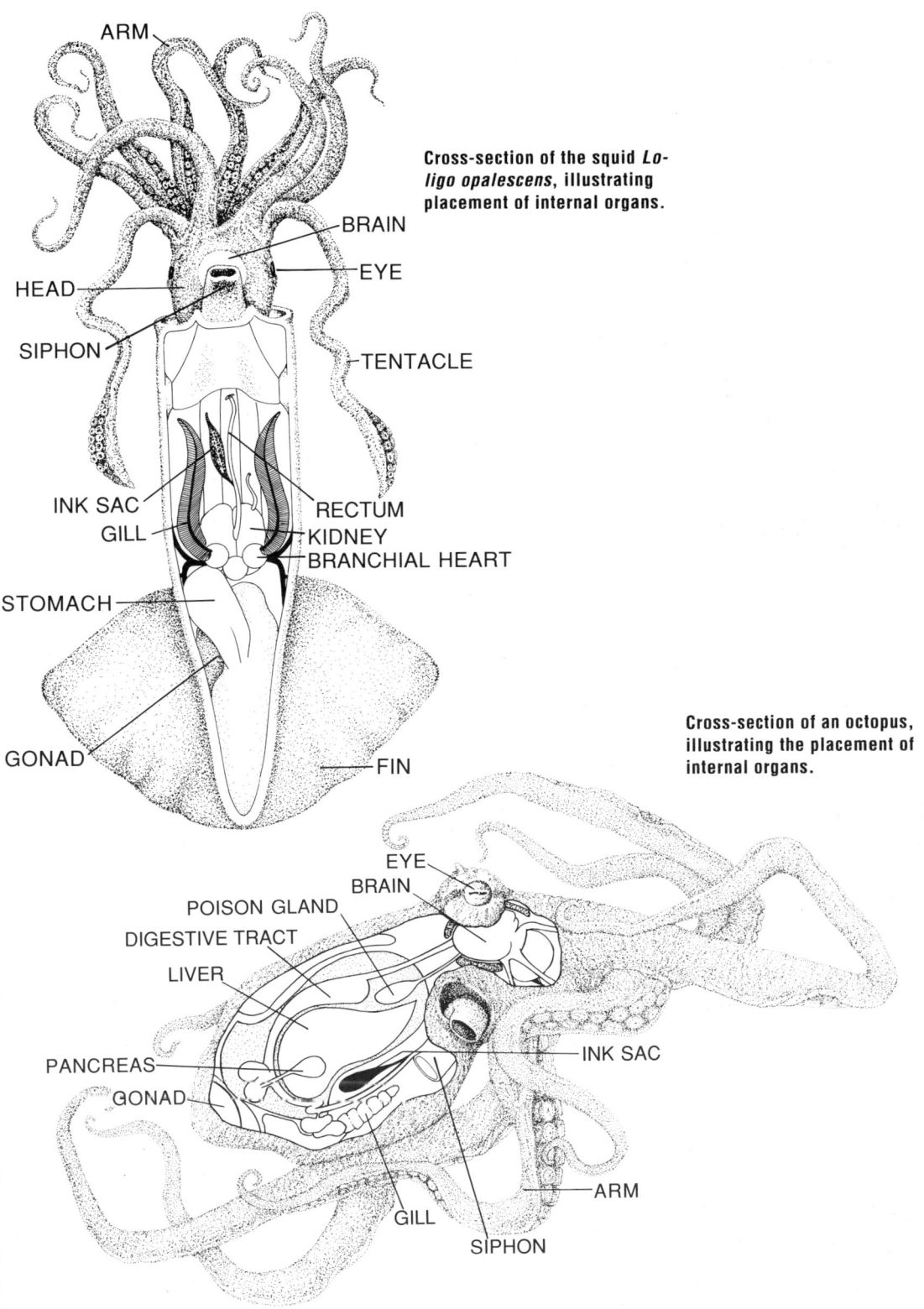

ARM

Cross-section of the squid *Lo-ligo opalescens*, illustrating placement of internal organs.

BRAIN

EYE

HEAD

SIPHON

TENTACLE

INK SAC

RECTUM

GILL

KIDNEY

BRANCHIAL HEART

STOMACH

GONAD

FIN

Cross-section of an octopus, illustrating the placement of internal organs.

EYE

BRAIN

POISON GLAND

DIGESTIVE TRACT

LIVER

PANCREAS

INK SAC

GONAD

ARM

GILL

SIPHON

95

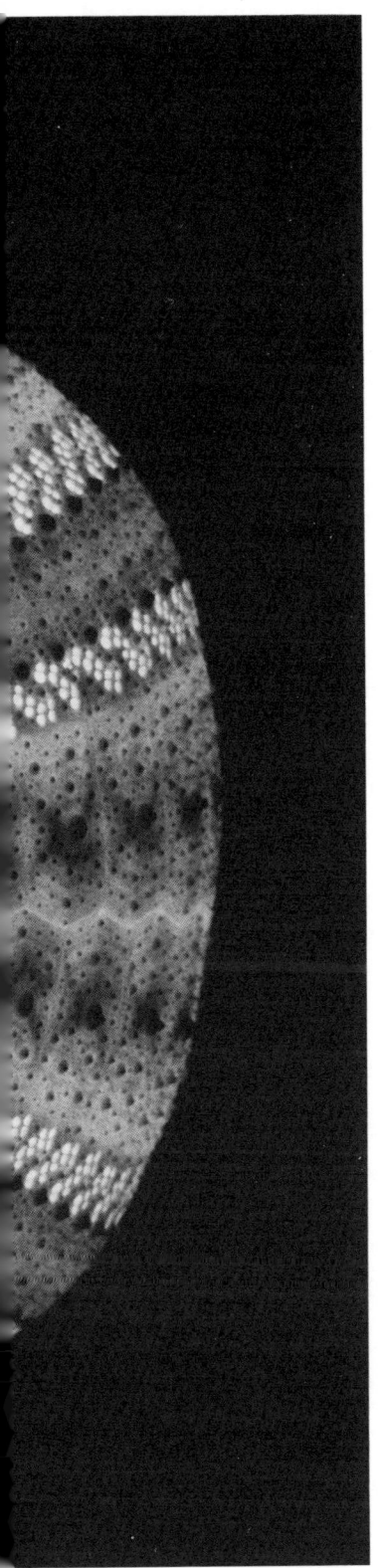

CHAPTER 8

Spiny-skinned Creatures

Most animals that move from place to place have a head end and a tail end. Typically, any sensory organs such as eyes or eye spots are at the front end. A mouth is usually at the head end, too. This is not so with starfish. A starfish's mouth is in the centre of its underside, and its eyes, such as they are, are at the tip of each arm. Even though starfish are fairly mobile, they are round, whether they have five arms or forty-four arms. And although they have a top and bottom, there are no sides. Sea urchins and sand dollars are the same.

The magnificent colours of some starfish inspire some people to take one or two home as souvenirs. But a sea star out of water for too long will die. It will then begin to decompose, or to be blunt, it will rot. The rotting process generates a powerful odour, which attracts attention from your neighbours, not to mention your neighbours' cat and dog, as well as rats you didn't know you had in your neighbourhood. The sea star's colours fade, and the drying carcass caves in where in life it was plump and rounded.

Starfish are best left on the beach, right side up, back where they were discovered in the first place.

Sunflower star (*Pycnopodia helianthoides*) is the largest sea star in the world and grows to 1 metre (39 inches) across. Adults typically have twenty to twenty-four arms; juveniles have fewer, accumulating more arms with increasing age. The animal's 15,000 tube feet give it speed and power, allowing the sunflower sea star to overcome and eat a huge variety of prey: sea urchins, scallops, worms, chitons, nudibranchs, snails, crabs and even other sea stars, such as the morning sun star.

The underside of a sea star's arm, showing the tube feet. Sea stars that live over sand or mud have pointed tube feet, and those that cling to rocks and hard surfaces have flattened suckerlike tube feet, as shown here.

Echinoderms are an abundant prominent group of marine animals. The name comes from the Greek word meaning "spiny skin." They are easily observed, since they tend to be large. They are exclusively marine and live very successfully in a great range of habitats. They are an ancient group, having evolved a number of distinct variations on a basic theme. Features unique to the echinoderms are tube feet and radial symmetry in the adult, replacing bilateral symmetry in the larva. Members of the group include the starfish, brittle star, sea urchin, sand dollar, sea lily and sea cucumber.

STARFISH

Although far from being the most common seashore animal, starfish, or sea stars, have the distinction of being one of the most noticed creatures of intertidal waters. They are biologically simple, but they are very much alive and share with all other living animals, including ourselves, the problems of survival. They are born and they move, eat, respire, reproduce, respond to their environment and die.

Sea stars first made their appearance in the oceans 425 to 500 million years ago. Today there are more than sixteen hundred species of starfish.

Like other echinoderms, sea stars are exclusively marine. Even though there are fresh-water shellfish, such as the fresh-water mussel, there are no fresh-water starfish. The greater number of species are deep-sea forms. An edged sea star (*Albatrossaster richardi*) was dredged from 6000 metres (19,685 feet) near the Cape Verde Islands. Fortunately, an abundant variety of these animals occur in shallower waters and can easily be seen at low tide. The West Coast is particularly rich in species.

A sea star is mainly arms, or rays, and most often there are five—though not necessarily. The beautiful rose star (*Crossaster papposus*) and sun star (genus *Solaster*) possess eight to twelve arms, whereas the sunflower star (*Pycnopodia helianthoides*) has up to twenty-four arms. The latter species is a giant of its kind measuring up to 90 centimetres (3 feet) across. The young sunflower star begins life with five or six rays, growing additional rays between the existing ones as it becomes older. The warm-water sunflower star (*Heliaster microbrachius*) develops up to forty-four arms. Consequently size, colour and number of arms may vary considerably within a species. Biologists must, therefore, use other criteria in identifying the various species of sea stars.

Starfish are clean animals. Unlike most bottom dwellers, the starfish does not serve as a home for algae or barnacles. If you look carefully at animals found among the rocks at low tide, you will see a high population density. Many organisms grow on top of each other; crabs, anemones, barnacles, corals, sponges and others all show amazing tolerance for invasion of their privacy. The tidy appearance of most starfish is due to hundreds of tiny pincerlike organs, called pedicellaria, which protect the breathing surface of the skin, crushing any animal or larvae that would settle on its back. The large and beautiful sunflower star has pedicellaria large enough to see under a magnifying glass if you happen to have both available at the same time. Biologists use the size, kind and pattern formation of pedicellaria as identification when the colour and size of the specimen are not definitive.

While the pedicellaria are crunching up some invaders, other debris is washed away in the mucus secreted by glands on the skin. In some species—the *Pteraster tesselatus*, for example—the mucus secreted is so thick and gelatinous that the animal has earned itself the name "slime star." This mucus is toxic to fishes. In an aquarium, many starfish cannot be displayed with fish, since the slime poisons the starfish's tank mates or clogs the gills of the fish.

Most sea stars are carnivorous and have voracious appetites, causing them to be viewed by the oyster farmer rather differently from the naturalist. Encircling the defenceless oyster with its many arms, the starfish pulls the oyster shell open a millimetre (a fraction of an inch), just enough to push its stomach into the shell. The starfish eats and digests the meat before retracting its stomach.

The Medusa of the starfish world, the basket star *(Gorgonocephalus eucnemis)*, is a fascinating deviation from the starfish norm. Its unusual shape is achieved by multiple branching of the five arms to create a waving basket of tendrils. Lacking pedicellaria and suckered tube feet, the basket star is thought to ensnare tiny fish and other planktonic creatures in its basket of arms.

Some sea stars are the housekeepers of the sea bottom, eating up dead and decaying sea life. Many are

Morning sun star (*Solaster dawsoni*) may range from orange to yellow or blue. The species typically has eight to fifteen arms, usually twelve, and grows to 30 centimetres (12 inches) across. It preys on other starfish, such as the vermilion star, blood star, leather star and sunflower star, with a marked preference for its close relative, the sun star (*Solaster stimpsoni*). It will even cannibalize its own kind.

Pink sea star (*Pisaster brevispinus*), when seen intertidally, often appears soft and flaccid, lacking the tough rigidity of its close cousin, the purple star. This indicates the species is less well adapted to intertidal life than *Pisaster ochraceus*. The pink sea star is an active predator on sand dollars and clams, and its presence in a sand dollar bed will cause the sand dollars within a metre's radius of the sea star to bury themselves quickly. The species may reach 60 centimetres (2 feet) across.

Spiny-skinned Creatures   99

also cannibals. Some species living in shallow mud bottoms consume mouthfuls of mud, digesting the organic matter contained in it.

The lifestyle of an average starfish does not require speed. The nearest candidate for long-distance runner is the sand star *(Luidia foliata)*, followed by the sunflower star, which has been clocked at 180 centimetres (6 feet) per minute. Most are slower. The principle of starfish locomotion is similar to our own. It is a system of leverage, or pushing the body forward. The thousands of tube feet found on the under surface of the arms are operated on a hydraulic system, the tube feet being hollow muscular tubes filled with water. When muscles at the body end of the tube feet contract, the foot extends, pushing the starfish forward. The muscular tube relaxes, the tube feet retract, and so it goes. It is a slow method of progression, but it works.

The tube feet are also capable of great suction; a pull of 3 to 4.5 kilograms (6½ to 10 pounds) has been measured on the sun star. The starfish uses suction, created by shortening the tube foot, to attach itself to rocks in pounding surf and to feed on bivalves. Suction is not the principal method of locomotion, however. How could the sea stars move over sand if it were? Contrary to common belief, starfish are able to right themselves if turned over, taking anything from a minute to an hour to complete the turn.

The water vascular system responsible for the operation of the tube feet has its partial intake on the upper surface of the animal. It appears as a small white plate, slightly off centre, and is called a madre-porite filter, or sieve plate. The water contained within the body serves as a medium for free-floating cells, carrying out much the same function as our blood.

One doesn't really think of a starfish as having a skeleton, but it does. The skeleton is composed of many small calcium plates, or ossicles, held together by a network of muscles to give the body rigidity and protection while remaining flexible. Species like the common intertidal purple star *(Pisaster ochraceus)* have very rigid bodies that are well protected from exposure to strong sunlight and wave action. Deep-water forms like the slime star tend to be more soft-bodied as the necessity for protection decreases.

Lacking a brain, starfish are still capable of co-ordi-

Purple star or ochre star (*Pisaster ochraceus*) is the most conspicuous of all sea stars on the West Coast. This species occurs in at least three colour phases: purple, ochre and brown-black. A tough exterior protects the purple star from desiccation as it forages high in the intertidal zone for mussels, barnacles, snails and limpets. Individuals may reach a diameter of 30 centimetres (12 inches). During the winter purple stars tend to migrate to the subtidal level. They appear to have few enemies, though sea gulls eat them and sea otters and seals may bite off an arm or two.

Six-rayed or brooding star (*Leptasterias hexactis*) is inconspicuous and drab in colour, though common and numerous. It is found under rocks intertidally and is full grown and large for its species at 9 centimetres (3½ inches). The species is also known as the brooding star because during the winter, from December to March, females raise their bodies and stand on "tiptoes," forming a cup and holding onto the eggs within the cup. If the female is detached and turned over, a cluster of golden-yellow eggs can be seen living in the cavity formed by the peculiar hunched posture. Only when the young have developed enough to cling to rocks does the female relax, a period of approximately forty days. Now flattened out, the female broods the young for another three weeks until they are 1 millimetre (a fraction of an inch) and fully formed.

Vermilion star (*Mediaster aequalis*). The pattern created by the armour of raised calcareous plates is most attractive in the sea star. In the northern parts of its range, from California to Alaska, it is fairly common intertidally and to depths in excess of 20 metres (65½ feet). The species grows to 17.5 centimetres (7 inches). It eats sea pens and tunicates.

Blood star (*Henricia leviuscula*), unlike many sea stars, lacks the tiny pincerlike organs, known as pedicellaria, that cover the dorsal surfaces of most other starfish. The species ranges from the low intertidal to deep water and reaches a diameter of 18 centimetres (8 inches). The species may be red, orange, yellow or almost white.

Look for this sea star where there are growths of sponges or bryozoans. The animal eats these as well as bacteria and small particles trapped in mucus on its upper side.

Brittle star (*Ophiopholis aculeata*). Serpent or brittle stars, like the sea lilies, are related to, yet distinct from, sea stars. Brittle stars are distinguished by their small size and five thin, snaky arms radiating from a distinct central disk. The group lacks pedicellaria, and although tube feet are present, they are not used in locomotion but for food gathering. Compared with sea stars, brittle stars move quickly, rapidly writhing their arms. Essentially, two pull and three push. Brittle stars are so named for their tendency to drop arms or portions thereof at the slightest provocation. Arms so automotized are quickly regenerated. Some brittle stars gather detritus and small organisms from the surface of mud, sand or rocks. Others wave their arms "overhead" to ensnare suspended particles from the water. Tube feet on the arms under the surface pass entrapped food particles and mucus "hand over hand" to the mouth. Unlike the sea stars, brittle stars are not able to extrude their stomachs out of their mouths.

Most brittle stars release eggs and sperm to the water with no brooding or parental care whatsoever. However, some brittle stars brood their eggs.

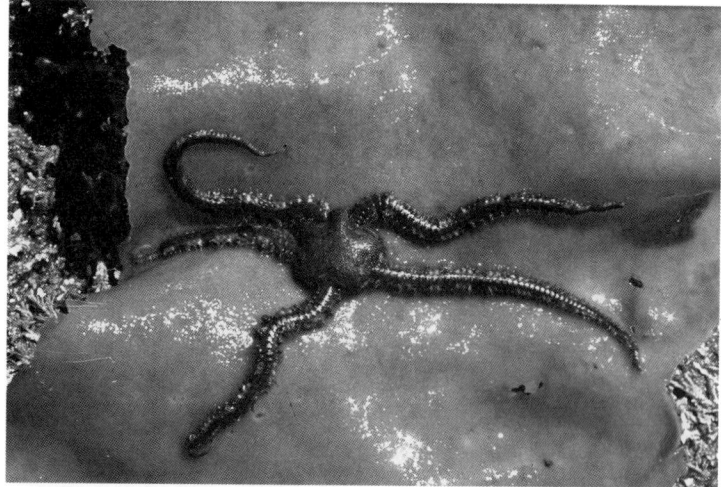

Bat star *(Patiria miniata)* is easily distinguished from the other common "web-footed" sea star, the leather star, by its rough, granular surface, which is very different from the leather smoothness of the latter. The bat star is a real scavenger, extending its stomach over a great range of plants and animals, dead or alive. Colour is extremely variable. The bat star differs also in being sexually ripe year round, whereas most sea stars have a definite annual reproductive season. Individuals may grow to 25 centimetres (10 inches) and can be found in tide pools, under rocks in the low intertidal and in deeper waters.

Leather star *(Dermasterias imbricata)*. The calcareous skeletal plates of the leather star are so small and deeply embedded in the "skin" as to give this species a distinctive smooth and slick exterior—hence the common name "leather star." The approach of a leather star will cause an otherwise stationary sea anemone to release its foothold and move off. Leather stars feed largely on sea anemones but will also eat sea urchins and sea cucumbers. These are all animals of rocky shores. The leather star reaches a diameter of 25 centimetres (10 inches), prefers rocky shores and is frequently encountered in the low intertidal zone. A peculiar garlic smell is associated with the leather star.

nated movement and have the ability to "sense" qualities of their environment. A cluster of simple eyes at the tip of each ray, called the "red spot," can distinguish between light and dark. How, then, do starfish find their food? Nerve endings originating from a central nerve ring appear to report chemical substances and vibrations in the surrounding area. The tube feet at the ends of the arms seem to be particularly sensitive.

Parenthood is a casual affair, so much so that parents never meet, much less set up house. The sexes are separate and, except by dissection, cannot be distinguished as male and female. The reproductive organs are located at the base of, and extend into, the arms. Eggs and sperm are released into the water through minute pores, and fertilization is left to chance. If fertilization occurs, a free-swimming larval form results and soon settles on the sea bottom. At this stage many young starfish fall prey to other organisms.

Two West Coast species, the six-rayed star *(Leptasterias hexactis)* and the blood star *(Henricia leviuscula)*, retain their young in brood pouches around the mouth until they assume adult shape. If a young starfish manages to survive the initial hazards of life on the ocean floor, it can look forward to a life span of about four years. There are exceptions, though, such as the purple star *(Pisaster ochraceus)*, believed to be the longest lived at twenty years.

Fortunately, what would be considered mortal injury to other animals does not necessarily mean death for sea stars. They possess an exceptional ability to replace lost parts. All living things have some capacity for repair or replacement of lost parts, but the starfish, deficient in other areas, surpasses higher animals here. Species of *Linkia* can regenerate a whole new starfish from a single arm. Most other species require that a good portion of the central disk remain intact; even so, it is a remarkable feat. The mechanism involved has puzzled and fascinated medical science for obvious reasons.

Successful with a minimum of biological equipment, the sea stars established themselves long ago and have survived. They are unchallenged; for the most part their only enemies are themselves, occasionally sea birds and sea otters and, of course, people.

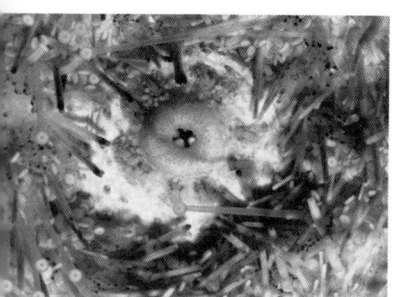

The underside of a sea urchin, showing its mouth and teeth. Note the tube feet between the spines. These pass entrapped bits of seaweed to the urchin's mouth.

Aristotle's lantern is not a lantern at all but is the complicated assemblage of teeth and "bones" that makes up the sea urchin's feeding mechanism. The forty "bones" are operated by sixty muscles. In the living urchin only the tips of the five teeth can be seen surrounding its mouth, which is located in the middle of its underside. Teeth, which become worn with use, grow continuously.

SEA URCHINS (ECHINOIDEA)

Sea urchins are first cousins to the sea stars, though the close relationship may not at first be apparent in a healthy, live urchin. The animal bears more resemblance to a pincushion than it does to its many-armed relative. However, one has only to examine the urchin's skeleton, or test, as it is called, to discover a pattern of five radiating divisions. Unlike the sea star, with its scattered bits and pieces of skeleton, the urchin's is composed of many fused plates creating a single, somewhat depressed hollow ball, open in the bottom centre. Small holes for the passage of tube feet and knobs for the articulation of the movable spines are arranged in definite radiating patterns over the test, making it, with its great beauty and symmetry, one of nature's works of art.

At first glance it appears that nature overdid it in the defence department. A few spines are definitely a deterrent to some would-be predators, but does the urchin really need as many spines as were provided? In some species, the abundance of spines may be related to the animals' feeding habits. Lacking the powerful, flexible arms of the sea stars to hold and subdue live prey, urchins must exploit other, less demanding food sources, such as bits of seaweed, which by chance become caught on and between the sharp spines. Seaweed so caught is passed "hand over hand" to the mouth by the tube feet, which are seen as long threads extending from between the spines. Because the tube feet occur over the whole urchin body rather than exclusively on the under surface, as in sea stars, urchins are able to use the whole body as an incidental food trap.

Like sea stars, sea urchins have pedicellaria, which have undergone some modifications. Rather than a pair of jaws, there are maybe three or even five jaws, which are elevated on stalks, enabling the tiny jaws to reach out and bite. Depending on the species of urchin, the pedicellaria may be adapted to special assignments. Some are designed simply to remove unwanted squatters, others to hold small animals for delivery to the mouth. Still others contain poison and act with the spines as a protective device against predators. Pedicellaria are also used for catching food; thousands of fish and invertebrates release thousands of larvae, most of which are looking for a place to settle.

In addition to this passive acceptance of nourishment, urchins are not above moving about in search of food. They will graze on many attached plants and animals and, in short, are very much opportunity feeders.

Because the sea urchin exploits a large range of animal and vegetable material as food, it must be able to chew the food enough to be handled by the animal's digestive tract. In the sea urchin's middle bottom is a mouth. Hidden within is a wonderful device known as Aristotle's lantern, which is used not to light the way but as a multipurpose chewing machine. The apparatus is so named because Aristotle described the five-tooth chewer as resembling an ancient five-sided lantern. The complete apparatus, a wonder of biological engineering, is composed of forty separate ossicles, or bones, arranged in a circular fashion and moved by sixty separate muscles. The latter move the teeth out, in or sideways, or rotate them in turn like an auger. The teeth can be made to bite, scrape, chew or drill. Encrusting animals such as bryozoans and calcareous tube worms are defenceless against the urchin's teeth.

In addition, by lifting and depressing the lantern structure, the urchin maintains water circulation within the body cavities. Thus, the lantern serves as a bellows to aid respiration as well.

Sea urchins move over the sea bottom using their tube feet. Some species also use their bottom spines as stilts. Movements are seldom extensive, and many urchins return to a home site for a part of each day. Other species on surf-pounded rocky coasts seldom move, preferring the security of excavated stony cups in which they sit. The small purple urchin (Strongylocentrotus purpuratus) is commonly seen living this way. How the excavated depressions are made in the rocks is still a matter of conjecture. However, it is thought that over a long period of time the spines and teeth of many urchins grind out the holes.

If damaged, sea urchins are unable to respond with quite the tremendous regenerative powers as the starfish. Lost spines, tube feet and pedicellaria are replaced, and though damage to the test may heal over, it bears permanent disfigurement.

Sea urchins attract much interest in many parts of

Green urchin (Strongylocentrotus droebachiensis), which has short, blunt spines, grows to 8 centimetres (3 inches) and ranges from Washington north. It also occurs in the Atlantic. This is the West Coast species of tide pools and protected rocky shores. It is thought to feed largely on bits of seaweed. Most sea urchins have a life span of four to eight years.

Purple urchin (*Strongylocentrotus purpuratus*) occurs intertidally and deeper in wave-washed exposed areas, where it frequently inhabits depressions scraped out of rocks. The spines are dense and short. Young specimens may be greenish; adults are purple and reach a size of 8 centimetres (3¼ inches). Range is from Alaska to California. Large purple urchins may be ten years old. Some are thought to be thirty years old.

Also shown on the left is the long-spined giant red urchin (*Strongylocentrotus franciscanus*). It grows to 14.5 centimetres (5¾ inches), the spines to at least 5 centimetres (2 inches), and ranges from California to Alaska. This giant red urchin occurs in a number of colour phases from pink to purple. It is by far the largest of West Coast sea urchins and is often seen in large aggregations, forming extensive subtidal mats in areas protected yet subject to good flushing action, like rocky channels. Some may be exposed at an extreme low tide.

the world, not because of their unique shape or vivid colours but because of their delicious roe. The roe is actually the ripe gonads of the mature male or female urchin and is reputed to be at its sumptuous best in late fall and early winter. The yellow gonads are eaten raw or strained and then whipped to a creamy consistency to be eaten as a spread on toasted French bread. The giant red urchin is harvested for export. Urchins over 10 centimetres (4 inches) across the test are collected by divers and transferred to the packing plant, where the gonads are removed. The roe is then dipped into a weak alum solution, washed and packed. The price fetched for this delicacy on the Japanese market is substantial, perhaps in part because of the belief that sea urchin eggs enhance sexual prowess.

Those urchins whose lives are not cut short to satisfy the gastronomic pleasures of urchin-egg eaters may look forward to a fairly long, uncomplicated and unharassed existence. As a free-swimming larva or a just-settled juvenile, the sea urchin undergoes substantial predation by starfish and many species of fish. But once established and well-spined, only the sea otter, some sea stars and humans are a real threat.

SAND DOLLARS

Sand dollars, or sand cookies, are essentially flattened sea urchins. The long, sharp spines and tube feet of the urchin are abbreviated to create a lush velvet coat on the animal's exterior. Like the sea urchin, the sand dollar has a single skeleton or test. It is reinforced internally by small pillars or braces to prevent crushing. The grey or white sand dollars found on the beach bearing a distinct flower pattern of five petals are not live sand dollars but the skeletons of once-living specimens. Common to the West Coast, the living sand dollars (*Dendraster excentricus*) are dark purple to almost black and will seldom be seen lying exposed on the sand at low tide; more likely they will be hidden just below the surface.

When covered with water, the sand dollar digs one-third of its body into the sand so that the top two-thirds is vertical and exposed to the tidal flow. Thus exposed, it becomes a food catcher trapping particles of food in the mucus between the short spines. But how does the food reach the sand dollar's mouth, which is located in the middle of its lower surface? Among the spines are cilia, or short hairs, which constantly beat, and in so doing direct the mucus in trails towards the mouth. Like small tributaries joining a larger stream, the mucous tracts are consolidated again and again into five main tracts. The tracts are directed so that food particles landing on the topside of the sand dollar, can be transported to the outside edge, over it and along the underside to the mouth. Like the sea urchin, the sand dollar has an Aristotle's lantern of five teeth. The structure is much smaller in the sand dollar, as would be expected, considering the small size of the food it eats.

As the tide recedes, the sand dollar falls flat and, using its movable spines, digs itself into the sand until the seas return again.

Sand dollars are animals of clean sand beaches and will often be seen in aggregations of many individuals. When one comes upon a group as the tide is receding, it recalls an overturned cookie box, the cookies lying helter-skelter, not quite a heap. Most likely sand dollars congregate in groups to increase the odds of reproducing more sand dollars.

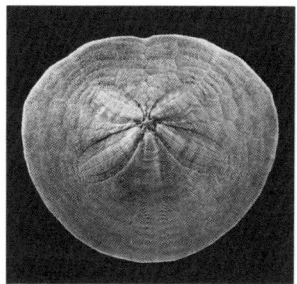

Sand dollar test, the skeleton of the once-living sand dollar. Starry flounders and pink sea stars, which are common in the same sandy habitat, are predators of sand dollars.

Like sea urchins and starfish, the sand dollars are separate sexually yet do not copulate or engage in any courtship activity. Most likely the release of eggs or sperm by one individual in the group stimulates others nearby to follow suit. By the simultaneous releasing of sex products, the probability that egg and sperm will meet is greatly enhanced.

SEA CUCUMBERS (HOLOTHUROIDEA)

A sea cucumber may look like a slug's cousin or a worm's fat sister with a bad case of warts, but it is not related to either. Oddly enough, a sea cucumber is akin to a transformed sea urchin, as if the urchin has been squeezed around the middle, so that the body is stretched upward. Spines and pedicellaria are absent, the skin is softened and the skeleton eliminated except for a few scattered ossicles. The resulting on-end sausage has five strips of tube feet running from top to bottom, the same five sets of tube feet that the sea urchin inherited from the ancestor of the five-rayed sea star.

For a sea cucumber, standing on its mouth would be awkward, if not impossible for a body without a firm skeleton, so the stretched-out, softened-up, despined creature falls over on one side. Three of five strips of

Giant red cucumber *(Stichopus californicus)* grows to 40
centimetres (15¾ inches).
This is an enormous cucumber adorned with many soft,
fleshy horns on the back of its
body. Although the body is
typically red to dark maroon,
the feeding mops are white. It
can be found on rocky shores
protected from strong wave
action. It feeds on organic detritus and small organisms.
When attacked by its enemy
the sunflower sea star, it
loops like an oversize inch-
worm in its attempt to escape.

tube feet contact the surface and are retained as functional tube feet. In most cases the two upper strips of
tube feet, no longer needed, either disappear or are
transformed into decorative warts and fleshy horns.
Thus lies before you the standard sea cucumber model,
such as the giant red cucumber *(Stichopus californicus)*.
Other species display variations and are described later.

The sea cucumber bears only a faint resemblance to
its symmetrically round kin, having acquired an elemen-
tary bilateral symmetry in having the mouth at one end,
the anus at the other and a definite directional orientation.

Some sea cucumbers are the underwater dust mops
of the sea, gathering up bits of detritus, small crustacea, protozoans and larval forms, using the sticky, mop-
like ends of specially adapted tube feet that surround
the mouth. When a mop becomes loaded with food, it
goes into the sea cucumber's mouth and is licked clean
in the same way a child licks jam off sticky fingers. Many
species are able to retract the feeding tentacles completely into the mouth cavity when not in use. Such is
the case with the giant red cucumber and the small milk-
coloured sea gherkin *(Eupentacta quinquesemita)* found
intertidally clinging to the undersides of rocks. The burrowing red cucumber *(Cucumaria miniata)* of the same
habitat does not always completely retract its crown of
bright orange-red mops and, consequently, is easily
identified by their presence.

Not all sea cucumbers live in a rocky shore habitat.
Some, mainly tropical species, live on a sand bottom,
and others, such as the burrowing sea cucumber *(Lepto-
synapta clarki)*, live a true burrowing existence buried in
the sand or mud. The latter species has little need of
the elaborate feeding mops of surface-living cucumbers,
since it gains nourishment by digesting organic material taken in as the cucumber eats its way through the
substrate in the manner of the earthworm. Like the
earthworm, the true burrower performs a valuable
function in shifting and mixing the substrate.

Most surface-living cucumbers do not breathe
through the skin, mouth or tube feet but through the
anus. Water is drawn internally through a many-
branched network known as the respiratory tree. Gas
exchange takes place across the respiratory membrane.
Each treeful of water is forcibly exhaled before the next

load of water is taken in. Knowing this, an observer can discern which end of the cucumber is fore or aft, depending upon which is doing the breathing.

A sea cucumber's anus is an ample opening and an inviting refuge for assorted small crabs, flatworms and at least one species of fish. The pearl fish (members of Carapidae), a small eel-like fish of warmer seas, has a most interesting association: the juvenile enters the sea cucumber head first through the anus, breaks through the respiratory membrane and establishes itself in the host's body cavity, where it feeds on the cucumber's gonads. The cucumber is not left sexless, since the gonads can be regenerated. As an adult, the pearl fish is free living, but it continues to recognize a sea cucumber's cloaca as a good place to hide when not searching for food. It has been found that the fish does not attach itself with any permanence to a particular animal but will enter any sea cucumber if need be.

A sea cucumber's freeloaders can be more correctly termed parasites, or commensals, depending on the degree of charity unwillingly exacted from its host. The adult pearl fish could be termed a commensal, since it gains protection through its association with the sea cucumber but not to the host's detriment. The juvenile pearl fish is a parasite, since it benefits at the expense of the sea cucumber.

For the crab or any other creature taking advantage of the sea cucumber's cloaca for shelter and protection, it must come as a surprise when the cucumber eviscerates and suddenly the freeloaders are thrust into the cruel world outside. When the sea cucumber eviscerates, it throws out its complete guts, usually via the anus. Once thrown out, the internal organs are not retracted. A revolting development; the organs become disassociated and creep about for a time on their own. New organs will regenerate in as little as nine days for tropical species or as long as three months in colder-water species such as the giant red cucumber. The cucumber appears to manage quite well without its organs while they are being replaced.

Why do sea cucumbers eviscerate? Not all species are prone to evisceration, though a great number are. In those that do eviscerate, it is occasionally in response to unfavourable conditions such as a rise or fall in water temperature or foul water, or for no apparent reason at

Red burrowing cucumber (*Cucumaria miniata*) grows to 20 centimetres (8 inches). The lowest intertidal zone from the Aleutians to southern California may reveal this cucumber secreted under overhangs or in crevices. The body is dark reddish. The most outstanding feature is a crown of bright red-orange feeding tentacles when the animal is submerged and feeding.

all. It appears that the giant red cucumbers eject their viscera every fall, since all specimens collected at this time lack internal organs. Perhaps they are undergoing an annual eviction of unwanted tenants or simply a renewal.

More frequently, evisceration is in response to attack. A crab, lobster, or fish attempting to molest the cucumber becomes entangled in the sticky entrails. The cucumber detaches itself from the whole and departs.

Still, evisceration is a radical and inefficient response to predation. Some sea cucumbers, generally tropical species, have developed cuverian tubercles as a means of reducing the losses resulting from evisceration. The tubercles are tough, sticky, stretchy threads that can be ejected out the anus to entrap the enemy. The threads are lost once ejected, as the internal organs would have been had they been ejected instead. By the sacrifice of the tubercles, the viscera are saved.

A sea cucumber simply cannot go throwing its guts out every time the shadow of a predator looms nearby. Surely, attacks or potential attacks must be a daily occurrence in the life of such a soft-bodied, slow-moving creature. One is left to assume the animal must be in possession of a secret weapon. Correct! Like the nudibranchs, many sea cucumbers have a nasty-tasting skin, the result of a poison known as holothurin.

South Sea Islanders and peoples of the Indian Ocean have long been acquainted with holothurin and its powers, and mashed or chopped sea cucumber is the main ingredient of one fishing technique. A reef lagoon or other suitable area is poisoned by the mash; the affected fish are then easily netted as they come to the surface. Sea cucumber poison is not dangerous to people unless injected into the bloodstream. Since the cucumber bears no spines, this is not likely to happen.

People are major predators on adult sea cucumber. The animals are gathered as food, known in the marketplace as *bêche-de-mer* or by the Malayan term *trepang*. Cucumbers are collected, eviscerated, boiled to desalt and destroy the poison, then smoked, dried and chopped. Sea cucumbers thus prepared are used in making soups and stews, becoming transparent and gelatinous when cooked and reconstituted. It is unfortunate that trepang has not found favour with Western palates, since it is nutritious, low in calories and high in protein (between 50 and 60 per cent). China remains the largest market, importing hundreds of tons.

Sea cucumbers are thought to have a life span of five to eight years. Reproductive activity occurs annually and most likely involves aggregations of animals, as in other invertebrates that shed their eggs and sperm directly into the water. (Most species are separate sexually, yet a few hermaphrodites occur.) Some brooding behaviour takes place, as in the armoured sea cucumber *(Psolus chitonoides)*. Heavy plates covering the cucumber's back are raised up, forming small cradle-caves for the developing cucumbers. Like their cousins the sea urchins and sea stars, sea cucumbers have bilaterally symmetrical larvae.

Sea cucumbers are among the exclusive fraternity of invertebrate animals able to survive successfully in very deep water. At 4000 metres (13,123 feet) in depth, 50 per cent of the living organisms are sea cucumbers. A species known as *Myriotrochus brunni* was dredged from 10 200 metres (33,465 feet) from the Philippine Trench. The how and why of such survival, living where others cannot on bacteria and nematode worms, remains largely a mystery.

Feeding "mops" (mouth) of giant red cucumber.

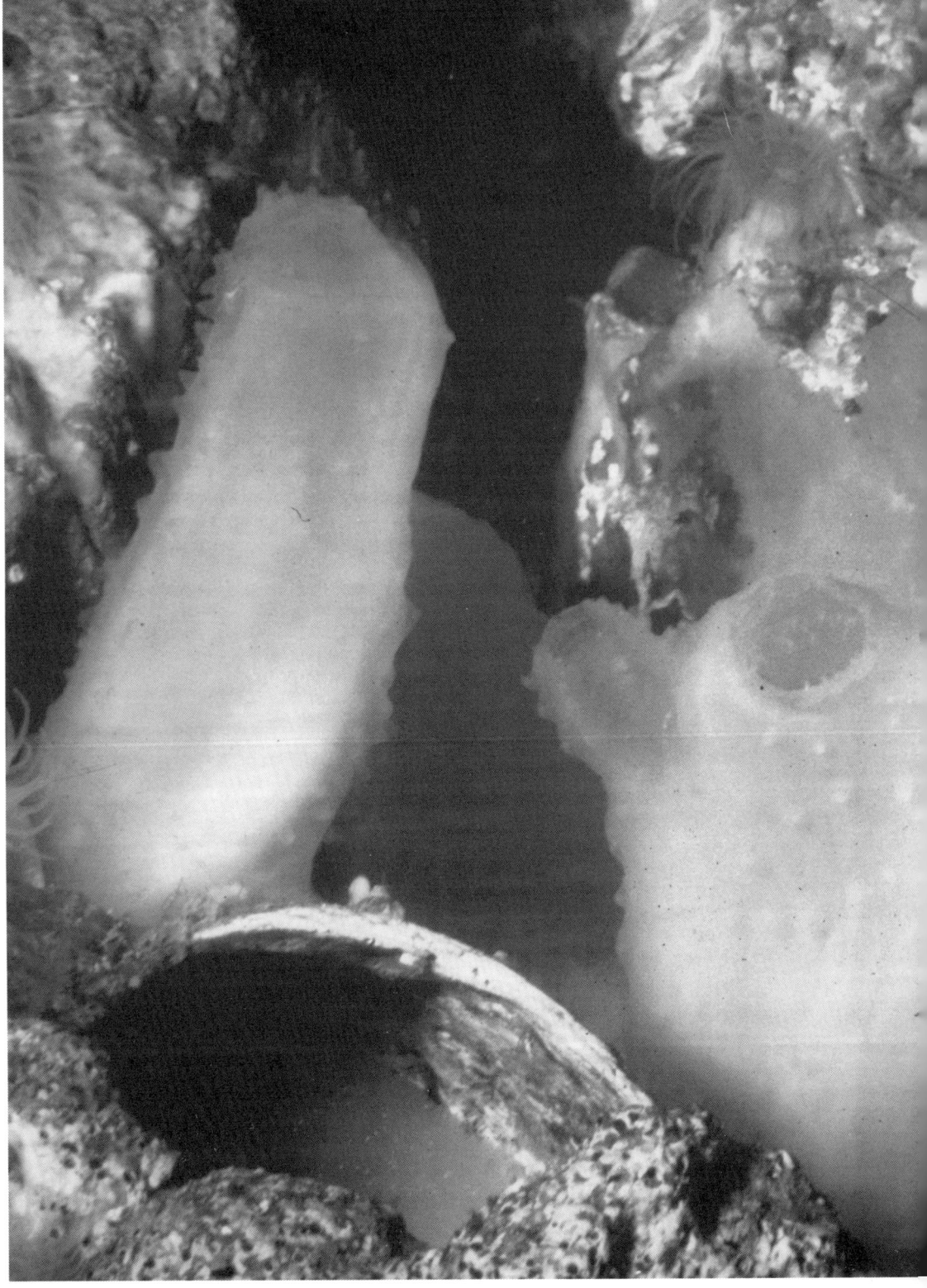

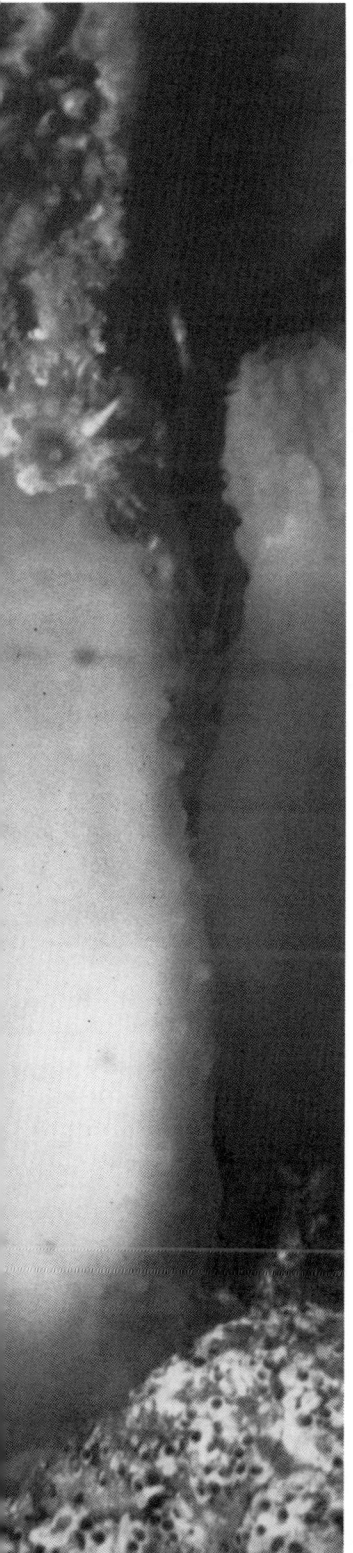

CHAPTER 9

Sea Squirts

Sea squirts are out there in the low intertidal, but they are small and hard to find. It takes patience and a real desire to want to see one. Like anything unfamiliar, the first one is the hardest to find. Floating buoys and anchor chains or dock pilings can be a real treasure chest of marine animals. I recommend starting there to look for your first sea squirt.

Warty sea squirt (*Pyura haustor*) has a thick, wrinkled tunic, which may be covered with debris and settling organisms. It occurs on docks, pilings and rocky shores, on calm bays, in rough water, on kelp holdfasts and in mussel beds. Look for a pair of distinctly red or pinkish rimmed siphons, on a small—to 5 centimetres (2 inches)—brown, warty animal.

Sea squirts are so named because they squirt water when disturbed. Shaped like a potato and ranging in size from a pea to a large spud, they are typically found attached to some firm substrate and range from the intertidal to great depths. In solitary forms, a pair of spouts, or siphons, distinguishes the sea squirt from other sessile organisms.

Sea squirts appear to mimic the lifestyle of sponges, lowliest of all many-celled animals. Both kinds of animals live out their adult lives in one location, and both depend on the bountiful sea to deliver food and oxygen. Yet despite such appearances, the sea squirt and the sponge are poles apart. In fact, the sea squirt is more closely related to humans than it is to a sponge.

Ascidians, as sea squirts are also known, belong to the phylum Chordata, of which frogs, fishes, whales, giraffes, mice and humans are also members. Most chordates have a backbone, but this alone is not enough to ensure identification. All chordates share three characteristics at least at some time in their life cycles: a notochord, a hollow nerve cord and pharyngeal clefts. The characteristics are mentioned not to confuse the reader but rather to help clarify the terribly difficult concept that a sea squirt is more closely related phylogenetically (racially) to vertebrates than to the sessile invertebrates it closely resembles.

To appreciate this strange relationship between backboned animals and sea squirts, it is necessary to look at the larval sea squirt.

Newly hatched sea squirts develop into free-swimming larvae, but not the simple ciliated larvae one would expect. Rather, the larva is a tadpolelike form complete with a notochord, which acts as a stiffening rod down the back providing skeletal support. A hollow nerve chord is located above the notochord and is expanded to a simple bulbous brain at its front end. Associated with the "brain" is a light-sensitive eye and a balancing organ. Pharyngeal slits open from the throat region to an atrium (passage). In more advanced chordates, these three primitive structures are replaced by more sophisticated structures: backbone (vertebrae), central nervous system and gill slits, respectively. As a tadpole, the sea squirt appears to be destined towards a bright future as some sort of vertebrate animal. However, after a brief career as a tadpole, the expected future is not realized. Instead, the larva glues itself head first to a substrate and begins to metamorphose into a stumpy adult. The notochord, nerve cord and tail are absorbed, and a saclike body with two siphons develops. The specialized sense organs of the larva disappear in the adult and are replaced with a ganglion and a few nerves to the internal organs.

As a sessile, limbless creature, the squirt has no means by which it can search for and select food items. It must, therefore, depend on small bits of food to be delivered via incoming water currents. Plankton and detritus suspended in the water are drawn into the body cavity through one of the two top spouts. The incurrent spout is termed the mouth pore, the excurrent is the atrial pore. A sievelike structure known as the branchial basket is suspended in the body cavity, and it is through this structure that all incoming water must pass. Mucous sheets lining the branchial basket trap food, which is then rolled into a rope of mucus and food that is fed into the sea squirt's digestive tract for digestion. The branchial basket is well supplied with blood vessels and acts as a respiratory organ to absorb oxygen, as well as a food trap.

Nutrients and oxygen are delivered to the various cells of the sea squirt by a most peculiar circulatory system. A simple, tubular heart pumps blood first in one direction and then, after a brief pause, in the opposite direction. A careful and patient observer should be able to watch this activity taking place in the transparent sea squirt *(Corella willmeriana)*, since it has an almost transparent body.

Sea squirts are also known by the term "tunicate" for the tunic, or coat, that covers the animal. The tunic may vary from soft jelly to a hairy covering or a plastic smooth exterior, depending on the species. The secreted tunicin that makes up the tunic is unusual in that it is a kind of cellulose and is not known to occur to any degree in any other multicelled animal except in the tubes of some sessile hemichordates.

Some species of tunicates occur not as individuals but as colonies of animals connected together by branching structures called stolons. Still others are embedded in a common matrix or tunic and are known as compound tunicates or compound ascidians. The latter often occur as encrusting jelly masses and are difficult to

The peanut sea squirt (*Styela gibbsii*) is wrinkled, hairy and peanut shaped, attached to the underside of rocks or pilings by a short stalk. It ranges from central California to British Columbia and grows to 7.5 centimetres (3 inches).

Small transparent sea squirt (*Corella willmeriana*) is a beautiful little animal, only 5 centimetres (2 inches) high. It is glassy clear and colourless, so a keen observer can see the animal's internal organs and possibly eggs in the atrial pocket. The sea squirt will be found in calm, clean water attached to a firm substrate such as a float.

Sea Squirts   115

distinguish from sponges and byrozoans in their encrusting forms. The smooth, jelly feel of the compound ascidian's matrix is generally distinctive.

Colonial and compound ascidians reproduce mainly by budding off new individuals from the parent without the aid of eggs or sperm. Solitary tunicates are nearly always hermaphroditic and shed eggs and sperm into the water to be fertilized there.

A free-swimming "tadpole larva" of a tunicate (sea squirt).

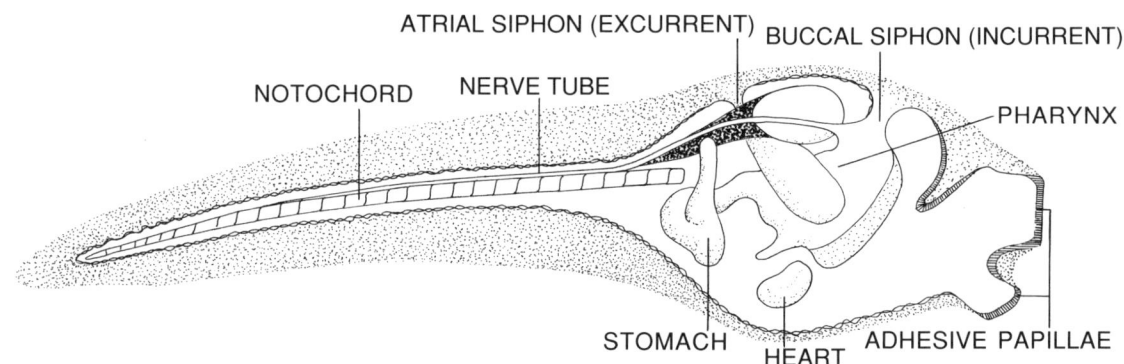

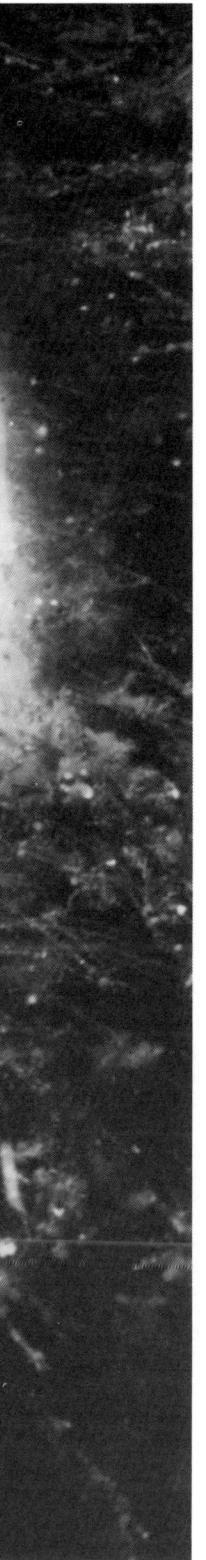

CHAPTER 10

Fishes

For many people, the word "fish" conjures up an image of a large grey carcass laid out on a bed of crushed ice at the fish market. For other people, "fish" connotes one of many small, brilliantly coloured swimming jewels in a tropical fish tank.

There are thousands upon thousands of different kinds of fish, and most are less than 30 centimetres (1 foot) long. Many fish swim all the time, but many don't. Many hop, slither or make short darts. These are the fish that spend time on the ocean bottom, living among rocks or in the underwater forests of seaweeds. These are the kinds of fish you are most likely to find in the intertidal: slimy little clingfish, stuck on the underside of a rock, or fantastically camouflaged sculpins darting about in a tide pool.

Even though there are many kinds of fish and the variations and adaptations at first appear complicated, it is well worth the effort to understand what makes a fish a fish. Fish are amazing animals.

There is no need to be afraid of a fish. Just keep your fingers out of its mouth, because I never met a fish that couldn't bite. If threatened, just about any animal will bite. Be comforted that the fish is not interested in eating you, and besides, its teeth are usually adapted to hold onto bite size pieces. Most fish cannot bite a chunk out of something large, and so their prey is limited to the size of their mouths.

There is a notable exception to the above description, and that is the shark. The shark's razor-sharp teeth enable it to eat anything, no matter how big or small, because they can slice chunks off an object too large to swallow whole. Imagine how this opens up the menu selection. This is not to suggest that you should be afraid of sharks but that you should be very wary of their teeth. Even small sharks like the dogfish can bend head to tail and therefore can easily bite a hand holding onto its tail. For this reason, be very careful when releasing an unwanted dogfish from a fishing line.

ABOUT FISHES

The lives of fishes can be likened to those of birds. The great albatross lives at sea and is strong and powerful. The diminutive hummingbird flits from blossom to blossom sucking nectar. A grouse scratches in the gravel and weed for seeds and berries. The falcon dives to kill its warm-blooded prey. Some birds thrive in the arctic tundra; others are of the forest, shore or plain. Each species, in adapting to its habitat, its food source and its enemies, has evolved habits and physical attributes that aid in its survival. Some are fast; others are slow. Some fly; others walk. Some are timid; others are aggressive.

In an underwater world of mountains, meadows and deserts, the same diversity exists in the lives of fishes because many of the same pressures to adapt and evolve in the air are also present in the ocean, although the sea is a different and infinitely more stable environment. To quote Pierre Teilhard de Chardin, fishes are "an assemblage of monstrous complexity."

There are some 45,000 species of vertebrates—animals with backbones. This figure includes all the fishes, amphibians, reptiles, birds and mammals. A little less than half the total vertebrate species are fishes. Since their beginnings, 450 million years ago, more than 20,000 species survive today.

Just about wherever there is water (over 70 per cent of the Earth's surface) there are fish—from the intertidal to depths of at least 9000 metres (29,528 feet), from the warm tropics to the sub-zero Arctic.

How does one describe a group of animals so diverse, in which the smallest member is an 11-millimetre (½-inch) goby *(Pandaka pygmaca)* and the largest is a 15-metre (50-foot) shark *(Rhincodon typus)* and in which body shape ranges from flatfish to sea horse? Perhaps the best place to begin is by asking, "What is a fish?"

A fish is a cold-blooded (poikilothermic) aquatic vertebrate. This means it is unable to regulate its own body temperature, as a bird or mammal does, and assumes the temperature of its surroundings. Usually it breathes by means of gills, has scales and possesses paired fins. A fish is mobile and has a head, a definite brain and elaborate sense organs. For every generalization, there is an exception, and therefore one expects to find fish species that live for long periods out of water, do not have scales, lack fins and so on. Space here does not allow for more than an acknowledgement that such exceptions do exist, and in great numbers.

Like all other living organisms, the fishes must respire, eat, avoid their enemies and reproduce themselves if they are to survive as a group. The development of an internal skeletal support with a movable yet rigid backbone and, in most cases, paired fins, has endowed the fishes with a mobility far superior to that of the invertebrates. Concomitant with their structural sophistication has come a greatly improved nervous and sensory system.

Open-ocean, or pelagic, fishes have streamlined, well-muscled bodies adapted to swift motion, with the tail, or caudal, fin providing the main thrust. Most highly developed in this respect are the tuna-type fishes, with their almost-rigid bodies and lunate (crescent-shaped) tails joined to the body by a narrow caudal peduncle (stem). Body muscles move the tail from side to side so rapidly it fairly vibrates, sending species such as the blue-fin tuna *(Thunnus thynnus)* through the water at speeds of 70.5 kilometres per hour (44 miles per hour). The power and hydrodynamic design necessary for such speed can be better appreciated when one considers that water is 800 times denser than air. Salmon (genus *Oncorhynchus)* can sustain 16 kilometres per hour (10 miles per hour) for short periods—about twenty seconds—and maintain speeds of 12.75 kilometres per hour (8 miles per hour) when cruising.

The majority of fishes, however, are not adapted for speed but for manoeuvrability. This can be expected, since most species are adapted to life in connection with the shore or bottom. In such cases fins are generally large, and the pelvic or ventral fins have moved forward, almost under the head, as in the sculpins (family Cottidae) and rockfishes (genus *Sebastes).* Many shore and bottom dwellers use their pectoral fins like props or long fingers to pull themselves over the sea bottom.

Bottom and shore dwellers cannot be considered without mention of the swim bladder, an organ that acts like a float. Divers will immediately understand the principle if they think of it as a "buoyancy compensator vest." When divers enter the water, the buoyancy of their tanks and wet suits is cancelled out by their weight belts, and they are able to descend as deeply as they choose. If

Eulachon (*Thaleichthys pacificus*). This fish played an important role in early Indian economy on the West Coast as food and as a source of fat for local use and barter. The fish is unique in that the rendered fat is solid at ordinary temperatures. In early spring great concentrations of eulachon come into fresh water to spawn, attracting large numbers of predators: dogfish, sturgeon, halibut, cod, whales, seals, sea lions and gulls.

Tide pool sculpin (*Oigocottus maculosus*), as the common name suggests, is frequently found in tide pools of rocky shores. Colour is variable, from red-brown or red to green on the upper surface; it is paler on the belly. The species grows to 9 centimetres (3½ inches) and ranges from northern California to the Bering Sea. The tide pool sculpin blends so well with its environment that it takes great patience to spot one, though many may be present. They move in quick dashes, a good strategy to keep from being seen and eaten by some predator.

they wish to remain at, say, 9.15 metres (30 feet) they must keep moving their fins to prevent sinking further. If they have buoyancy compensator vests, they simply blow enough air into the vests to allow themselves to remain buoyant at that point. More air will cause divers to rise, less to sink.

Fish with swim bladders have the equivalent of a buoyancy compensator built into their internal structures as an off-pocketing of the gut. If the swim bladder is connected to the gut, as in the salmon, the fish rises to the surface to gulp and swallow air. If the swim bladder has become closed off, as in the rockfishes, the fish is able to produce or reabsorb its own swim bladder gases. For species that hang on rock faces, the swim bladder allows the fish to maintain its position in aquatic space without expending energy to do so. Often when anglers are fishing for rockfish (sometimes called rock cod), the fish they bring to the surface has bulging eyes and its swim bladder has been forced out through the mouth. The grotesque condition of the fish is caused by rapidly decreasing hydrostatic pressure as it is brought to the surface. Unable to reabsorb the quickly expanding gases in the swim bladder, it literally balloons within the fish's body.

Because sharks do not have swim bladders, they must remain on the move to sustain their position in the water. When they stop moving, they sink. Some fishes that do not move up and down within the water column but remain on the bottom, such as flatfish and some sculpins, poachers, eelpouts and gunnels, have greatly reduced swim bladders or have lost the organ entirely.

To some extent body shape and body colouring are related. For example, most pelagic species, such as the salmon, herring and tuna, are silver bodied with countershading. That is, the fish's back is darker so that when seen from above by predators or potential prey, the animal will blend with the darkness of the water below. The belly is light so that when viewed from beneath the fish will blend with the silver reflection of the water's surface and the sky.

Bottom dwellers, such as flatfish, are generally mottled or shaded to resemble the substrate they inhabit and are usually coloured only on the exposed side, being white beneath. The sand sole (*Psettichthys melanostic-*

Cross-section of a fish.

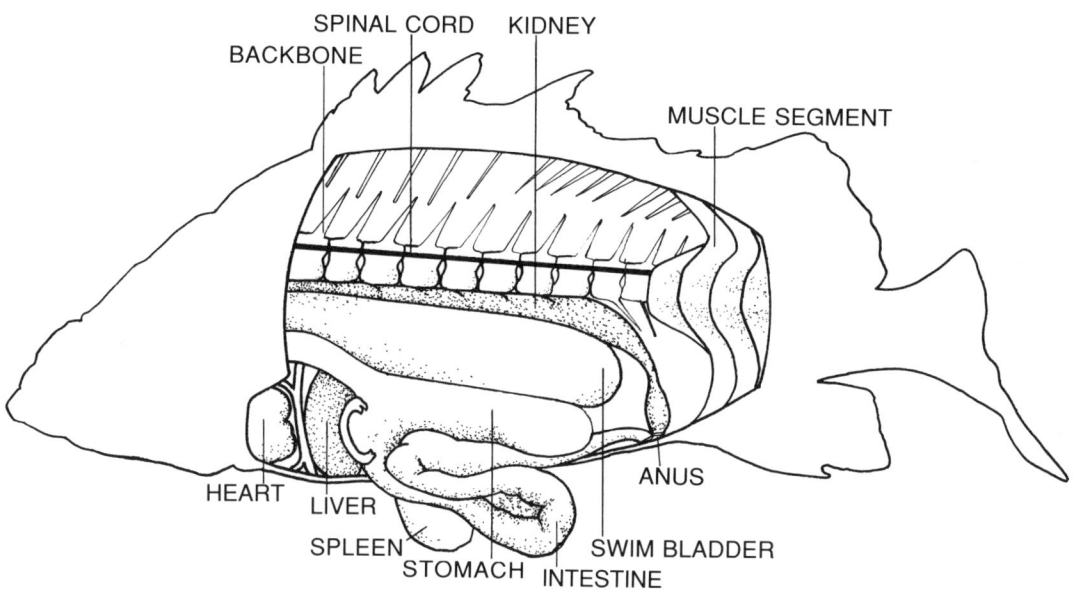

SPINAL CORD   KIDNEY

BACKBONE

MUSCLE SEGMENT

HEART   LIVER

SPLEEN   STOMACH   INTESTINE

SWIM BLADDER

ANUS

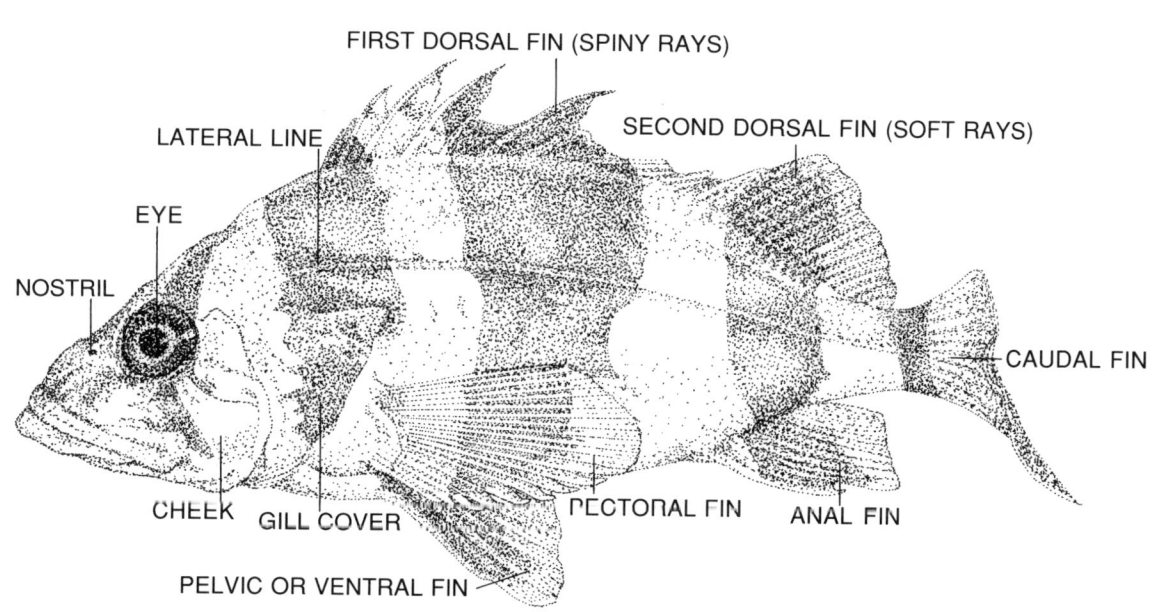

FIRST DORSAL FIN (SPINY RAYS)

LATERAL LINE

SECOND DORSAL FIN (SOFT RAYS)

EYE

NOSTRIL

CAUDAL FIN

CHEEK   GILL COVER

PECTORAL FIN   ANAL FIN

PELVIC OR VENTRAL FIN

Crescent gunnel (*Pholis laeta*) is an eel-like fish, but it is not an eel. It is common under rocks, in tide pools and to depths of 55 to 73 metres (180 to 240 feet). It is named for the many crescent-shaped markings on the dorsal fin against a body ground colour of yellowish-green. The scientific name is a Greek and Latin combination meaning "one who lies joyfully in wait." The species grows to 25 centimetres (10 inches) and ranges from northern California to the Bering Sea and through the Aleutians.

The red Irish lord (*Hemilepidotus hemilepidotus*) is predominantly red with spotting and mottling of brown, white and black. Adults feed on crabs, barnacles and mussels and are known to bury themselves in the sand. Red Irish lords grow to 51 centimetres (20 inches) and range from California to Alaska.

tus) and many other flatfishes perfectly mimic the sandy bottoms where they are found.

Flatfishes are actually capable of darkening, lightening or mottling, by expanding or contracting pigment in their chromatophores. Thus, on a light, sandy bottom they will be a rather uniform pale colour, whereas on a gravelly bottom they will be mottled with areas of light and dark. Experiments indicate that the stimulus to alter the colour pattern is visual, since blinded flatfishes are unable to alter their colour.

Rocky-bottom species often exhibit cryptic (protective) colouration as well. The red Irish lord (*Hemilepidotus hemilepidotus*) is nearly invisible until it moves—so perfectly does it mimic the algae- and sponge-encrusted rocks where it lives. Stripes, spots and bars of all descriptions and colours occur on sedentary, rocky-bottom fishes, and not all colouration is for camouflage. In some territorial fishes, such as the tiger or blackbanded rockfish (*Sebastes nigrocinctus*), the bold colouration, as in this species' bold stripes, acts as a warning signal to other members of the species.

Colouration cannot be considered without mention of body shape as well. Frequently the fish's body contour and posture are as much responsible for its cryptic nature as is its colour. The flatfishes are an obvious example. These fishes begin life as regular, vertically oriented fishes, taking to the bottom and falling over to one side after a brief larval stage. The eye of what becomes the lower side migrates to the upper side so that even though the fish is forever on one side, both eyes are on top. However, one gill remains on the bottom, and the mouth develops a curious, distorted shape. Skates and rays, however, are truly flattened from top to bottom and have two eyes, naturally on top, and both gills on the underside.

The pipefish (*Syngnathus griseolineatus*) and tubesnouts (*Aulorhynchus flavidus*), with their long, thin tubular bodies, look for all the world like bits of sticks or grass. Neither species is endowed with much speed, which is unnecessary, since their peculiar body shape provides effective concealment from enemies.

All manner of bumps, lumps, projections and spines obliterate the outlines of many fish species. A marvelous example here is the amazing headgear of the decorated warbonnet (*Chirolophis decoratus*). Sitting among the rocks, with its head adorned with what look like little Christmas trees, the decorated warbonnet is well camouflaged from its enemies yet able to dart out and seize prey before being recognized.

Spiny parts of many species, particularly the rockfishes, are for defence rather than camouflage. When threatened, the rockfishes throw their sharp dorsal spines forward to repel potential predators. In fact, the rockfishes are related to the most deadly venomous of all fishes, the tropical stonefish *(Synanceja horrida)*. The poison glands in the stonefish are associated with each of the thirteen dorsal spines. Should an unwary beachwalker step on one of these, it is death within ninety minutes unless antivenom serum is injected. Other rockfish, though incapable of such spectacular defence activity, can still give a nasty jab and result in an even nastier infection.

Fishes constantly secrete mucous slime from special cells and glands in the skin. This coating protects the animal from abrasion, infection and some parasites, as well as discouraging settling organisms such as barnacles, sponges and hydroids.

Under the slime coat there are usually thin, flexible scales arranged like overlapping shingles. Scales vary widely in shape and size; gunnels have such small scales that they appear scaleless, and in poachers the scales are modified to hard, bony plates.

A fish's scales are not shed as the animal grows but are added to, growing larger with the fish. For this reason, where seasonal fluctuation of temperature affects a fish's metabolism, as on the West Coast, growth rings like tree rings are evident on the scales and can be used to determine the age of the specimen (growth is rapid in summer, slower in winter). The ages of fish lacking scales may be calculated by using growth marks on the ear bones or vertebrae, but this method necessitates killing the fish. Scale counts are harmless to the fish, since lost scales are soon replaced with new ones.

Most fish "breathe" water; that is, they absorb oxygen dissolved from water. The water passes in through the mouth, over the gill arches in the pharynx and out, under the gill cover, or operculum. Although fish appear to be gulping water, two internal pumps, slightly out of phase, maintain a steady stream over the gills. Because volume for volume water contains less dissolved oxygen

C-O sole eyes and mouth showing how in true flatfishes the lower eye migrates to the upper surface but the mouth remains where it is when the young fish swims upright.

Decorated warbonnet (*Chirolophis decoratus*) is subtidal, occurring at depths from 18 to 91 metres (60 to 300 feet) and is larger and more robust than the other gunnels described here. The bizarre head ornament serves to camouflage the fish from its enemies and its potential prey as it sits among growths of algae and hydroids. Warbonnets grow to 42 centimetres (16½ inches) and are marked in tan and brown. The species occurs in inshore waters from Washington to the Aleutians.

than air, the fish gill is, of necessity, up to four times more efficient in taking up available oxygen than the human lung. Many fine filaments on the (usually) five gill arches per side present a massive surface area to the flow of water. In the mackerel, total gill surface area is ten times that of the body surface. "Rakers" on the inside of each arch prevent suspended matter from fouling the gills.

Fishes are sensitive creatures. They see with monocular vision, and some are able to discriminate a broad spectrum of colour. The capacity for colour vision is variable among species. Many deep-water forms have been found to lack cone cells—the cells responsible for colour perception—in the retina and therefore see only in shades of black and white. For many years it had been thought that sharks saw only in black and white. Experiments now show that sharks have cone cells in their retinas and are able to perceive some colours—yellow, for example. Fish lack true eyelids and have no tear glands. Nasal openings in the form of paired or single pits "smell" substances dissolved in water. (It is thought that the smell of a salmon's home stream is, in part, responsible for its ability to return sometimes thousands of kilometres to the waterway of its birth.) Fishes taste; that is, they perceive chemicals in contact, as opposed to those at a distance, as is the case with smell. Taste receptors are not limited to the mouth and tongue but may occur on the head and body surfaces.

Fishes produce and perceive sounds. They create noises in various ways—using muscles in connection with the swim bladder or by rubbing hard surfaces such as pharyngeal teeth and spines. The function of sound production in fish is not yet firmly established but is thought to be related to courtship or territorial activity, though this does not explain why the intriguing little grunt sculpin *(Rhamphocottus richardsoni)* makes its peculiar grunting noises when removed from the water.

Fishes have no outer ear, eardrum or Eustachian tube but are able to perceive sound vibrations in water. A more important function of the auditory apparatus is that of balance, just as the human ear is the centre of equilibrium.

Fishes have yet another sensory link with the environment—the lateral line that runs the length of both sides of the animal's body approximately at the middle

Pacific tomcod (*Microgadus proximus*). Many West Coast fishes are referred to as "cods," yet only a few species of true codfishes (family Gadidae) occur here. The Pacific tomcod is one of these fishes. True codfishes are distinguished by having three separate dorsal fins. The Pacific tomcod grows to 30 centimetres (12 inches) and is well regarded as a table fish. The species ranges from central California to the Bering Sea.

and branches into three more lines in the head region. Its basic component is a neuromast, a group of sensory cells. Minute pores in the fish's skin open into tiny internal pits, each supplied with hairlike projections that are extraordinarily sensitive to pressure waves and may be sensitive to electrical fields and other environmental fluctuations. Each cell group is amply supplied with nerves connecting the lateral line with the brain. The sensory function of the lateral line appears to serve as a means of "touch at a distance" by registering pressure changes in the water. If you have ever watched a large tank of aquarium fishes being fed, all animals converging en masse on the food presented, you have probably noticed that the fish never bump into each other or into objects in the tank. Perhaps this is because of the lateral line.

The same mechanism is at work in "schooling" fishes, when up to a million individuals, depending on the species, are massed together. Yet the school is more than a crowd of fish. Individual members are oriented in the same direction, travelling at the same speed, maintaining virtually equal space between them. The school moves as a single unit, all members turning or closing ranks in unison. There is no leader. Individual fish of a school are generally of equal size because larger individuals are faster than smaller fish of the same species. Consequently, schooling fishes will tend to sort themselves according to size and speed.

What influences a fish to join others of the same species to form a school? What are the advantages of a school? The answer to the first question is still unknown, but it has been established that schooling behaviour begins very early in the life of a young fish; tentative attempts at grouping and orientation have been observed in newly hatched fry.

Schooling is an effective adaptation, but why this is so is still largely unanswered. Two thousand schooling marine species and another two thousand schooling fresh-water species, including both primitive and advanced forms, must find a definite advantage in the school formation. The school may serve to confuse predators, yet some predatory species school. The school may enhance reproduction if eggs and sperm are shed directly to the water, but some schools are composed of only one sex. Schooling may facilitate food

The grunt sculpin (*Rhamphocottus richardsoni*) is so named for the peculiar grunting noises it makes when removed from water. It pulls itself over rocky bottoms using fingerlike pectoral fins. A long, pointed snout pokes into crevices and between barnacles searching for small crustaceans and other organisms. Grunt sculpins grow to 7.5 centimetres (3 inches) and range from California to the Bering Sea.

Herring (*Clupea harengus pallasi*) are small—to 33 centimetres (13 inches)—unspecialized, schooling fish. They have no spines, no teeth on the jaws—no special features at all. Herring are abundant, and their importance to humans and the ocean's food chain is due in great part to this abundance. Herring are major food fodder for chinook and coho salmon, waterfowl, sea lions, seals, dogfish, lingcod and whales. Commercially, herring is fished for bait and reduction to oil and meal. There is also a smaller market for pickled or kippered herring and herring roe.

finding, yet only individuals on the outside of the school are able to search for it. Perhaps it is a means of conserving energy by employing group hydrodynamics; the exertion of each fish may be lessened because it can utilize the turbulence produced by the surrounding fish. Yet animals of the school's leading edge exert no more energy than would a solitary fish.

It follows that the greater sensory input of fishes over that of the invertebrates would necessitate a larger, more complex brain. Bony fishes are capable of learning and, through the use of memory, are able to develop relatively complex conditioned responses. However, it is difficult to find behaviour in fishes that can be unequivocally attributed to thought or reason.

Successful reproduction of a species is essential to its survival, and to this end the fishes have evolved some of the most fantastic and improbable means of reproduction imaginable. Space here will allow for only a few brief descriptions, and then only as they apply to species of the West Coast.

Fish are generally separated into two sexes; male produce sperm in testes, and females produce eggs in ovaries. With the exception of the sharks and skates and the live-bearing surfperches, brotulas and rockfish, most West Coast species lay eggs. Some, like the lemon sole *(Parophrys vetulus)*, release huge numbers of free-floating, or pelagic, eggs, and the number of eggs released increases with the size of the female. For example, a 30-centimetre (1-foot) female averages 150,000 eggs, and a 44-centimetre (17-inch) female averages nearly 2 million. Eggs hatch within ninety hours in California and take somewhat longer in northern regions. After a pelagic larval stage of six to eight weeks, the young soles settle and metamorphose into flatfish. Another flatfish, the halibut *(Hippoglossus stenolepis)* may release up to 2.7 million pelagic eggs but does so only after reaching maturity at ten or twelve years of age. Female halibut may grow to 213.5 kilograms (470 pounds) and live to thirty-five years, whereas males achieve only a fraction of this weight, at 18 kilograms (40 pounds) and live to twenty-five years.

Lingcod *(Ophiodon elongatus)* are somewhat less casual, the female depositing her half million eggs in rock crevices or under overhanging boulders. A large egg mass may be up to 61 centimetres (2 feet) long and weigh 13.5 kilograms (30 pounds). After depositing the eggs, the female departs, leaving the male lingcod to guard and fan the egg mass until hatching takes place about two months later, when his guarding activities cease. As in the halibut, females are larger than males. The female lingcod may reach 137 centimetres (4½ feet) and 45.4 kilograms (100 pounds), whereas the male is only about 91.5 centimetres (3 feet) and 11.3 kilograms (25 pounds).

Tubesnout (*Aulorhynchus flavidus*) males go one better. They not only guard eggs but build a nest as well. Tubesnouts are schoolers, disrupting their schooling behaviour briefly during the spawning season when males establish territories and build nests by winding seaweed together with strong, weblike strands extruded from the urogenital opening. Small bands of ripe females school above the nests, eventually depositing egg masses not in, but on, the nest. The males guard these nests until hatching occurs a few weeks later. In this species sexes are of similar size, about 165 millimetres (6.5 inches), and are thought to have an annual life cycle.

The somewhat similar looking pipefish (*Syngnathus griseolineatus*), relative of the sea horse (genus *Hippocampus*), exhibits a high degree of parental care. After an elaborate courtship ritual, a ripe female entwines her long, thin body around that of the male and deposits her fertilized eggs into the male's abdominal brood pouch, where the young remain until they are 19 millimetres (¾ inch) in length. Once free of the pouch, the juveniles are on their own.

These few examples give some idea of the enormous range of reproductive activity in fishes. Naturally, when eggs are laid in the thousands or millions, it can be expected that only a small percentage will survive to maturity. Many eggs are destroyed before hatching by storms, drying, fouling from fungus or sedimentation and predation by sea gulls, diving ducks, starfish, crabs and other fishes, to name but a few predators. If an embryo should survive to hatching, it faces uncounted terrors, mostly because of its small size. The hatchling becomes fair prey to any and all carnivorous animals that are bigger and faster than itself. Most larval fishes wait out the crucial period after hatching, as they gain in size and strength, among the shallow protected

Pacific halibut (*Hippoglossus stenolepis*). The halibut illustrated here was over 400 kilograms (880 pounds) before head and entrails were removed. It is the largest taken in recent years off the coast of British Columbia; a larger, 404-kilogram (890-pound) fish (dressed weight) was taken from the same area off the Queen Charlotte Islands in 1914. The specimen shown here was taken on a 19-kilometre (12-mile)-long set line with hooks at 5.5-metre (18-foot) intervals.

Female halibut are typically larger than males, with maximum recorded dressed weights of 216 kilograms (476 pounds) for females and 56 kilograms (123 pounds) for males. These are records, however, and most are proportionately smaller. Unlike most flatfishes, which are sedentary bottom dwellers, halibut are fast-moving, active predators that feed voraciously on other fish. When taken on hook and line, the halibut is a formidable fighting fish.

Bay pipefish (*Syngnathus griseolineatus*). Although the tubesnout and pipefish are similar in appearance, they are not related. The pipefish is distinguished by having a body wrapped in bony plates and by lacking pectoral fins.

The female pipefish transfers her eggs to the male's abdominal brood pouch. There the eggs remain and develop until hatching. An elaborate courtship display precedes mating.

Pipefish frequent eelgrass beds, wharves and shore pilings, where they can be seen moving in short, jerky thrusts, sucking small organisms into their mouths through a long, tubular beak. The species ranges from central California to southeastern Alaska and reaches a length of 33 centimetres (13 inches).

waters of eelgrass beds, sargassum weed and kelp beds. Because these areas are critical as fish nurseries, it is essential that they be preserved and protected not only from physical disturbance but from pollution as well.

Fishes have lifestyles just as we do, and often one is able to draw a fairly accurate picture of a particular fish's activities and behaviour just by knowing what clues to look for.

The kind of mouth a fish has can be very revealing. For example, fishes with whiskers, or barbels, on their snouts usually have an underslung mouth, like the poacher *(Agonus acipenserinus)* and sturgeon (genus *Acipenser*). It can be concluded from this feature that the fish is a bottom feeder, perhaps a scavenger rooting around muddy bottoms for things to eat. The barbels indicate that the fish may live where the water is dark, or turbid, and uses the barbels to sense food where it cannot be seen. This is how sturgeon living at the mouth of the Fraser River use their barbels.

A huge, wide mouth, accompanied by large pectoral fins, points to an opportunity feeder like many sculpins (Cottidae), often improperly referred to as bullheads, and lingcod *(Ophiodon elongatus)*. These fish are not fast predators but move in short bursts to snatch in their huge maws anything edible that comes along such as fish, crabs and shrimp. In very deep sea fish, the whole animal becomes mouth—an adaptation for capturing anything where food is scarce and meals are few and far between.

A small forward-positioned mouth on a streamlined body points to a fast, open-water predator such as the salmon (genus *Oncorhynchus*), or tuna. Here the mouth is armed with many fine, sharp teeth, efficient for capturing prey on the wing. For these fish, their best offence is speed; a large mouth, as in the lingcod, would only get in the way and slow them down by interfering with the streamlined design of their bodies.

Anchovy *(Engraulis mordax)* and herring (genus *Clupea)* are streamlined fish but have large, extensible mouths that can be dropped open, becoming great plankton scoops. The peculiar sharp nose, stubby body and fingerlike pectoral fins of the grunt sculpin *(Rhamphocottus richardsoni)* indicate that it is a slow-moving bottom species that feeds by poking in among barnacles and rocks for small worms and other invertebrates.

The position of a fish's eye provides yet another clue to possible lifestyles. Eyes positioned low and flush to the sides of the head point to a streamlined and therefore fast-moving species—again, as in the salmon. Eyes positioned high on the head, or even raised, as in the flatfishes or staghorn sculpin *(Leptocottus armatus)*, suggest that these are fish that burrow into the sand or mud to conceal their presence, with only their eyes roving over the sea floor for potential prey.

Examples are endless, and the reader is encouraged to watch fish as Sherlock Holmes would, looking for clues to where the suspect has been and what it has been doing.

FISHES WITHOUT JAWS (AGNATHA)

Two very unlovely aquatic vertebrates barely make the grade as fish. These are the hagfish and the lamprey (Cyclostomes or Agnatha) of which two species each occur on the West Coast. Superficially eel-like, with large, round, fleshy sucking mouths at one end and anuses at the other, these creatures exist as scavengers, parasites and predators on large fish. They have no jaws, no paired fins, no scales and, in the hagfish, no obvious eyes. There are gill pouches: six to fourteen in the hagfish, seven in the lamprey.

The hagfish attacks its victim from the inside, having entered through the mouth or anus, and literally eats it from the inside out. It first digests the internal organs, then the muscle tissues of the living or dead host fish. Eventually all that remains is the victim's empty skin. Cod, lingcod, flounder, salmon and dogfish are commonly infested with this pest. One dogfish skin was found to contain four hagfish, and since the mature hagfish may be over 63.5 centimetres (2 feet) long, hosting four of the creatures is no small feat. Not surprisingly, the hagfish is a serious nuisance to fishing operations.

The Pacific lamprey *(Lampetra tridentatus)* attacks its victim from the outside, not from the inside. Firmly attaching itself to a host fish's skin, the lamprey gradually rasps a hole through it with sharp, horny teeth. Once so attached, the lamprey is in perpetual association with food, sucking the host's body fluids and tissues.

After an undetermined number of years at sea as a parasite on larger fishes, the Pacific lamprey returns to

Lingcod (*Ophiodon elongatus*) are not cod at all but belong to the greenling family. They occur from Baja California to Alaska in shallow water and to depths of 230 fathoms in the southern part of their range, and generally within the upper 50 fathoms in British Columbia. Females grow faster and larger than males. A very large male seldom exceeds 11.5 kilograms (25 pounds), whereas females have been recorded in excess of 45 kilograms (99 pounds). Towards the end of November in British Columbia the male establishes a site within his territory where the female will lay her eggs. After the eggs are laid in December, the female departs, leaving the male to guard the egg mass until hatching in March.

Lingcod are fished commercially for the fresh fish trade, particularly for use in making high-quality fish and chips.

Pacific hagfish (*Eptatretus stouti*) is a very primitive fish. It has no jaws, no fins and no scales. It lives buried in muddy ooze until it smells a fish nearby. The prey may be dead or alive. The hagfish then wriggles towards its victim and enters it to eat it from the inside out. The hagfish exudes copious amounts of slime if handled, an effective offense against predators. It grows to 63.5 centimetres (25 inches) and ranges from California to Alaska.

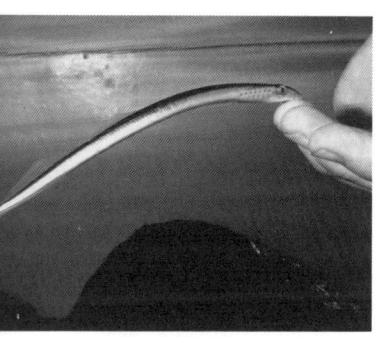

Juvenile Pacific lamprey (*Lampetra tidentatus*) temporarily attached to a person's finger to show how this animal would attach itself to a host fish.

a fresh-water stream sometime between July and October. The following spring, pairs of lamprey dig nests, spawn and die. Within two to four weeks a new generation of potential parasites has hatched as larvae. These are carried downstream to pools, where they bury themselves in the mud. For five years the blind, wormlike larvae live here eating diatoms (algae) and other small organisms, until at around 11 centimetres (4¼ inches), they transform into the adult form and go to sea.

SHARKS, SKATES AND RAYS (CHONDRICHTHYES)

The old adage that a few bad apples spoil the barrel is certainly applicable to the sharks. The fear and loathing that this group inspires rests on the very undesirable behaviour of a few. Of the approximately 280 species alive today, only a dozen or so eat humans, a number are considered threatening if provoked, and the rest are either too small or lack the kinds of jaws and teeth needed to inflict serious damage on human swimmers. Contrary to popular belief, human-eating sharks do not deliberately seek out human prey but, like so many "opportunity feeders," will take whatever is available, people included. Shark behaviour, particularly attacking behaviour, is of a highly unpredictable nature. Attacks are not limited to a particular location, time, season or other condition. However, statistics indicate that sharks prefer to take human prey in waters warmer than 20° C (68° F). (Perhaps this is because more human bathers will occur at any one time in warmer water than in cold.) Yet even this generalization is not without exception. The great white shark (*Carcharodon carcharias*), most feared man-eater of all, has been known to attack swimmers in waters of 13° C (55° F) off the coast of California. This species is wide-ranging and is thought to prefer temperate waters. In 1968, the carcass of a 5-metre (16½-foot) specimen was found on the beach of Graham Island in the Queen Charlotte Islands.

Only one problem species occurs with regularity in West Coast waters—the blue shark (*Prionace glauca*). Ranging from southern California to Alaska, the blue has not been proven to eat humans but is listed as potentially dangerous.

Sharks are an ancient group and, in many respects, a primitive one. Along with the skates and rays, which are essentially very flattened sharks, they are separated from most other fishes by having a skeleton not of bone but of cartilage. Hence the sharks, skates and rays are termed cartilaginous fishes. Unlike higher fishes, referred to as bony fishes, sharks and their relatives exhale water through multiple gill slits on either side of the head, as distinct from the single opening and cover plate (the operculum) in the bony fishes. Shark skin is not covered by the typical shinglelike scales of most other fishes but has embedded in its surface minute spines, giving the skin a very rough texture. These are known as denticles *(denticulus,* a "little tooth"), or placoid scales. Shark teeth are actually modified denticles, and they grow in several rows in the upper and lower jaws, the front row being functional and the back row being held in reserve for constant replacement. However, not all sharks use their teeth as a means for capturing prey. Both the whale shark *(Rhineodon typus)* and the 11-metre (36-foot) basking shark *(Cetorhinus maximum)* feed by straining plankton through sievelike structures on the gills known as gill-rakers. Water and plankton enter through the huge mouth. As the water passes over the gills, small organisms are retained on the inside of the gills, in the throat, and are swallowed.

Sharks do not spawn; that is, they do not release eggs and sperm into the water for fertilization. Both the sharks and rays copulate, the eggs being fertilized within the body of the female. In all male sharks and rays the inner edge of the pelvic fins is modified to form an elongate clasper, an erectile organ used to transfer sperm to the female during copulation. Eggs may be deposited in egg cases, as they are in most skates. Within the egg case, commonly referred to as a mermaid's purse, further development takes place, and the young eventually emerge from the case as miniatures of the adult. Or young may be born alive, as with most sharks, the eggs having hatched and developed within the mother's uterus.

In the spiny dogfish *(Squalus acanthius)* the time from copulation to birth is two years. At birth the pups are between 25 and 27.5 centimetres (10 to 11 inches), a good size considering the adult female is seldom more than 120 centimetres (4 feet) in length and males are smaller at just over 90 centimetres (3 feet). A single litter may range from two to seventeen young, but seven or eight is most usual.

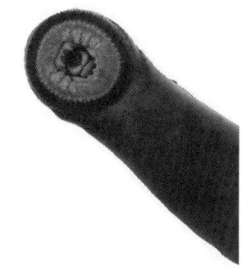

The lamprey's mouth has no jaws but is a funnel-like disk lined with horny "teeth." The lamprey attaches itself to a host fish and not only sucks blood from its victim but hitches a ride as well.

Pacific spiny dogfish shark (*Squalus acanthias*) is frequently caught on hook and line along the West Coast, since it will take almost any lure and will eat almost anything: squid, fish, lobsters, jellyfish and even algae. The characteristic pack feeding of this small shark is thought to have inspired the name "dogfish."

Big skate (*Raja binoculata*). The Latin name for this species means "two-eyed" for the two large spots, one on each pectoral fin. In British Columbia the big skate is common over muddy bottoms at moderate depths. Its range is from southern California to southeastern Alaska. The pectoral fins of this large—to 2.4 metres (8 feet)—skate are sold commercially as fresh fish. It is probably an omnivore, helping to explain why it often takes a lure and is caught by recreational fishermen.

Because most newborn sharks enter the world in an advanced state of development, fewer eggs are needed for survival of the species. This is in direct contrast to the herring, for example, which spawns thousands of free-floating pelagic eggs, which are exposed to all manner of physical disaster and predation from the very moment of fertilization.

Predators must be able to receive and interpret information from their environment quickly and accurately if they are to be effective hunters. Sharks are no exception. Vision, at least at close range, is good. Vibrations and pressure waves are perceived through the lateral line. Keenest of all is the sense of smell. A nose as we know it does not exist in the shark; rather there is a sac lined with sensory cells capable of responding to chemical "smells" in the water. Other small sensory sacs of the head region, known as the ampullae of Lorenzi, provide information about salinity, pressure, temperature and electrical current. Sharks are very much on top of what is going on within their world.

A species frequently taken in purse seines and on other fishing gear is the six-gill or mud shark *(Hexanchus griseus)*. It is a sluggish, deep-water species found in most temperate oceans of the world. The largest Pacific specimen recorded was 4.5 metres (15 feet) long.

By far the most commonly encountered shark on the West Coast is the spiny dogfish. It ranges in the eastern Pacific from Baja California to the Bering Sea but is most abundant between northern California and northern British Columbia. Two long spines, one in front of each dorsal fin, and the relatively small size of the species (between 90 and 160 centimetres [between 3 and 5 feet]) readily distinguish the dogfish from other local sharks. Carelessness when handling dogfish may result in a painful wound. On the back of each spine is a shallow groove containing venom. As the spine enters the victim's skin, the venom gland is damaged, releasing its fluid into the flesh.

Dogfish feed on a great variety of foods, principally herring, sand lance, smelt and shrimp. It is not surprising, then, that dogfish are attracted by the same bait and lures used in salmon fishing, since salmon feed on the same general diet. It is the general impression among sports fishermen that more dogfish occur in West Coast waters than all other species put together.

An exaggeration certainly, but the shark is abundant, and more so now than thirty years ago. This abundance is in part due to the collapse of what was a substantial commercial dogfish fishery. In the late 1930s, dogfish were harvested for the high vitamin A content of their livers. By 1950, liver oil imported from Japan and the advent of synthetic vitamins forced the fishery's closure. In addition, dogfish are a very long-lived species—to forty years—in contrast to a fish such as the salmon, which has a life span of four to seven years.

If one is able to see past the dogfish as a scourge and look at it simply as another fish, it soon becomes apparent that in design and motion the shark is near-perfection. It has a primitive body plan, little altered over the course of 200 million years, which serves the predator well.

A total of eleven species of shark occur on the West Coast.

A pair of skate egg cases, or "mermaid's purses," as they appeared shortly after being laid by a female big skate (*Raja binoculata*). These cases were 22.5 centimetres (9 inches) long. One case has been opened to reveal two eggs within.

RATFISH

A most peculiar animal to see and to classify is the ratfish. It belongs somewhere between the sharks and bony fishes. It has some sharklike characteristics, including a cartilage skeleton, paired claspers in the male for internal fertilization and eggs enclosed in horny capsules. Yet like bony fishes, ratfish have a gill plate cover (operculum) and a more advanced jaw structure.

Ratfish propel themselves not with the tail fin, as most fish do, but with broad sweeps of the huge pectoral fins, like a bird flapping its wings. A disproportionately large head with rabbitlike teeth, huge green eyes, a long, rat-like tail and a club-shaped appendage on the forehead of the male combine to create what can only be described as strange.

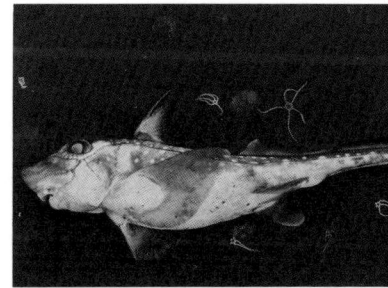

Ratfish (*Hydrolagus colliei*) is so named for its long, rodent-like tail. The fish swims by flapping its long pectoral fins.

BONY FISHES (TELEOSTOMI)

Two-thirds of all living fish species are classed together as bony fishes (the Teleostomi), as opposed to the hagfish and lampreys (the Agnatha) and the sharks, skates and rays (the Chondrichthyes). As would be expected, there exists within the class an enormous range of size, form and behaviour. Bony fishes range in size from 1 centimetre (½ inch) to 4.6 metres (15 feet). Most fall within the 2.5- to 30-centimetre (1- to 12-inch) range and therefore exploit a smaller living space than the sharks, for example. Well-developed jaws, greater and more precise mobility and a swim bladder have been responsible for much of this group's phenomenal success in the aquatic realm.

Bony fishes usually have true bone present in the skull, jaws or pectoral arch. There is one external gill opening covered by an operculum, and two sets of mobile paired fins, the pectorals (front or side pair) and pelvics (rear or ventral pair). Unlike the sharks, the bony fishes can twist, turn, spread or flatten the fins, allowing a much greater range and control of movement.

SALMON

Of the six species of Pacific salmon, five occur on the West Coast. (The Japanese cherry salmon (*Oncorhynchus masou*) is limited to the Japanese islands and nearby Asian mainland.) The different species vary in size, habits, life span and spawning time, yet all share a similar life history. Salmon begin life in fresh water, migrate to the sea and reach maturity there. As spawning time approaches, adults return to the fresh-water streams of their birth to spawn and then die. Fishes such as the Pacific lamprey and salmon, which return to fresh waters to spawn, are known as anadromous fishes.

The salmons are sleek and beautiful creatures. They are prized sports fish, make delicious eating and form the basis of an important commercial industry in both the United States and Canada.

ROCKFISH (SCORPAENIDAE)

There are more species of rockfish on the West Coast than any other fish family. In California, fifty-three species have been recorded; in British Columbia, thirty-five species. The family, Scorpaenidae, is a large one. Rockfish are distinguished by having eleven to seventeen long, heavy spines in the dorsal fin but lack the venom gland of their notorious cousin, the Indo-Pacific stonefish. Mouth and eyes are large, and many smaller spines occur on the head and gill-cover region. Rockfish of the genus *Sebastes* are ovoviviparous, meaning the eggs are retained and develop within the female's body and the young are born alive. Rockfish of the genus *Sebastolobus* are oviparous, meaning they lay eggs and do not give birth to live young.

Pink salmon (*Oncorhynchus gorbuscha*). One of five species of Pacific salmon common to the West Coast. All salmon hatch in fresh water, go to sea, where they grow and mature, and return to fresh water to spawn and die. Life spans range from two to seven years. Salmon are prized sports fish and valued commercial fish. They are also important prey fish for many other animals throughout their entire life cycle. The movements and concentrations of killer whales, seals and sea lions are frequently tied to the abundance of salmon along the coast.

All rockfish are edible, and some are delicious table fish.

Rockfish occur from shallow water to depths of 430 fathoms and are primarily bottom-dwelling fish. The large mobile fins and swim bladder afford them great mobility and stability in confined rocky areas where nonschooling species establish themselves in rocks and crevices.

The rockfishes are opportunity feeders, eating smaller fish, crustaceans, tunicates, jellyfish and squid. It must be mentioned that the name "rockcod" is a misnomer, since the rockfish are in no way related to the codfish.

SCULPINS (COTTIDAE)

This is a large group of fishes (some 300 species) occurring in a great variety of forms. Most are bottom dwellers of shallow waters. Sculpins are characterized by stout forward bodies tapering off to a much slimmer posterior portion. Eyes are usually large and positioned high on the head. Pectoral fins are typically fan shaped and large, with the pelvic fins positioned far forward.

POACHERS (AGONIDAE)

Poachers are generally small bottom fishes living at moderate depths. Some occur in deep water, a few in tide pools. The body is covered with bony plates that meet but do not overlap.

FLATFISHES

More than twenty-one different species of flatfishes occur off the West Coast. These are popularly termed sole, halibut, turbot and flounder. Most are bottom fishes, frequently in very deep water.

China rockfish (*Sebastes nebulosus*). A blue background with yellow and white side mottling make this one of the most attractive rockfishes. It grows to only 30 to 43 centimetres (12 to 17 inches) and ranges from central California to southeastern Alaska.

Recreational fishermen frequently catch rockfish near kelp beds, referring to their catch as rock cod.

Rockfish are slow-growing, long-lived fish and can be overfished in a short time. Care must be taken not to waste or overfish.

Shiner seaperch (*Cymatogaster aggregata*). Children delight in fishing for this shiner from wharves using mussel-baited hooks. The species has large scales and is always shiny, though the colour may range from bright silver to almost black for males during the breeding season, from April to July. Females are larger than males right from birth. Unlike most other fishes, the ambiotocids (shiners, seaperches and surfperches) are born alive as miniatures of the adults.

Pile seaperch (*Rhacochilus vacca*), as the common name indicates, frequents wharves and pilings. A favoured food is mussels, which are eaten whole, shells included, later to be voided in the feces in mucous casings. Pile seaperch can be distinguished from other seaperch of the same habitat by the higher, sail-like back half of the dorsal fin. Adults are dark grey or brown above, with silvery bellies and sides. Young pile seaperch have dark vertical bars. Adults grow to 44.25 centimetres (17½ inches) and range from northern Baja California to southeastern Alaska.

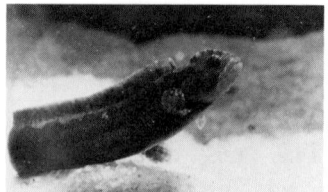

Cockscomb prickleback (*Anoplarchus purpurescens*) is the most common eel-like fish to be found under rocks at low tide along sheltered coasts. Its common name is inspired by the "cockscomb" on the head. During the winter a female will lay about 2700 eggs among the rocks. She guards these until hatching. Individual cockscombs travel little, remaining within a 15-metre (50-foot) area most of their lives. The species reaches a length of 20 centimetres (7¾ inches) and occurs in variable shades of grey, brown, olive and almost black. It ranges from California to Alaska.

Staghorn sculpin (*Leptocottus armatus*). A large—to 46 centimetres (18 inches)—scaleless sculpin, this species is abundant in shallow and protected sandy areas. It tends to bury itself in the sand, leaving only its bulbous eyes exposed. The common name "staghorn" refers to the antlerlike spine on the operculum (gill cover). Staghorn sculpins range from Baja California to the Gulf of Alaska.

Flathead clingfish (*Gobiesox meandricus*) are found under large rocks in the intertidal. The flathead clingfish remains attached to the underside of its rock by a sucker disk created by a modification of the pectoral and pelvic fins. The species looks like highly polished agate or marble, grows to 12.2 centimetres (5 inches) and feeds on small crustaceans and mollusks. The flathead clingfish ranges from southern California to southeast Alaska.

Threespine stickleback (*Gasterosteus aculeatus*) is a small—to 10 centimetres (4 inches)— silvery fish generously distributed throughout temperate northern areas in both fresh and salt water. In the latter, they occur in brackish harbours, in other coastal areas and far out at sea. Marine sticklebacks are considered to be anadromous fishes returning to fresh water to spawn.

Threespine sticklebacks are important forage for many larger predacious fishes such as trouts and pike.

The starry flounder (*Platichthys stellatus*) is a very common flatfish of shallow water and is easily distinguished by alternate dark and light bars on the dorsal and anal fins (side fins in the flatfishes). Although they grow to a good size—92 centimetres (3 feet) and 9 kilograms (20 pounds)—they are not highly regarded commercially. Like most flatfish, this species will lie in ambush buried in the sand with only its eyes above the substrate. You may well disturb such a flatfish when wading in shallow sandy water at low tide. Fear not, the fish eats shrimps, worms, clam siphons and small fish, never people.

Great sculpin (*Myoxocephalus polyacanthocephalus*). Sculpins (family Cottidae) are what many people commonly refer to as bullheads. They are in fact many different species in a large family of bottom-dwelling fishes known as sculpins. Most have stout forebodies, tapering dramatically towards the tail, large mouths and eyes, and large, fanlike pectoral fins. Most sculpins live in shallow water, and many occur in the intertidal zones. The large mouth and eyes, cryptic colouration and stationary habit indicate a fish that sits and waits to ambush prey. An opportunity feeder, the sculpin will typically pounce on any food item that will fit into its cavernous mouth. This is why large sculpins readily take a baited hook.

Plainfin midshipman (*Porichthys notatus*) was so named for the luminous organs dotting the fish's body in rows, reminiscent of the brass buttons on a midshipman's uniform. Although the species occurs to depths of 145 fathoms, adults are frequently encountered intertidally or in shallows in spring, when they excavate nests under rocks and then spawn. Males remain with the nest, guarding and caring for the eggs until hatching two to three weeks later. The midshipman is also known as the "singing fish" for the peculiar and loud noises it makes using the diaphragm across its swim bladder. Small fish and crustaceans form the fish's diet. The species grows to 38 centimetres (15 inches) and is thought to range from the Gulf of California to southeastern Alaska.

Striped seaperch (*Embiotoca lateralis*) is a strikingly coloured fish of the West Coast, with its bright copper background and electric blue horizontal stripes. Like other seaperches, the young are born alive and fully formed. Mating is preceded by an elaborate courtship, with the males approaching the females on their sides and quivering rapidly. The anal fin is modified as a copulatory organ. The species grows to 38 centimetres (15 inches) and ranges from northern Baja California to southeast Alaska.

Female kelp greenling (*Hexagrammos decagrammus*). The female of the species is pale in comparison with the male. Kelp greenlings, like their cousins the lingcod, are voracious feeders and have been reported to eat almost anything, including anemones. For this reason, perhaps, the species takes a lure and is often caught by casting from shore, when it is called a golden sea trout. Adults attain a length of 53 centimetres (21 inches).

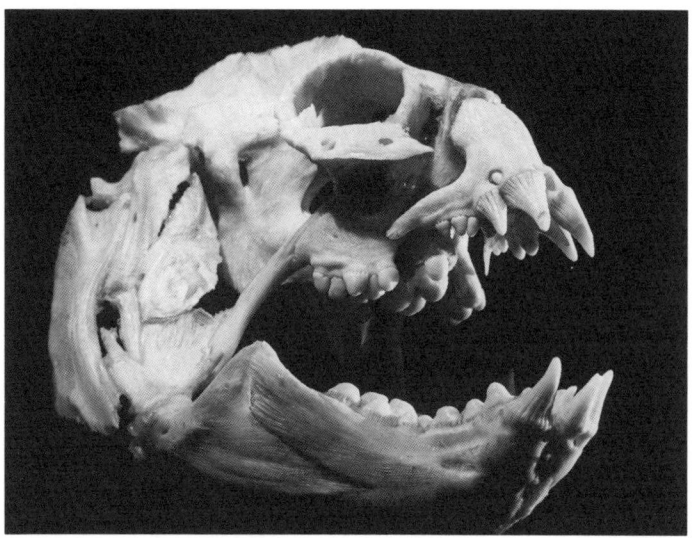

Wolf eel skull. Note the well-developed canine teeth used for crushing shelled invertebrates such as clams and crabs. Strong crushing molars farther back in the jaw grind food before it is swallowed. This skull was prepared from a 150-centimetre (5-foot) specimen.

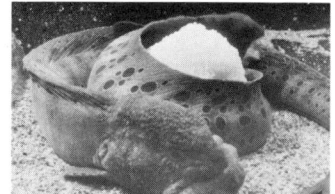

Wolf eel (*Anarrhichthys ocellatus*) is the largest of the eel-type fishes on the West Coast. No other fish is easily mistaken for this fierce-looking and powerful predator. Because they are sluggish swimmers, wolf eels generally attack only slow-moving prey such as crustaceans, urchins, clams, mussels and the occasional slow-moving bottom fish. Male and female pairs often inhabit the same rocky den in shallow to moderately deep water from southern California to the Gulf of Alaska. The female encircles the egg mass and maintains a water current over the eggs by undulating her long dorsal fin. The male wolf eel has a proportionately larger head and thicker lips than the female. Length is to 2.5 metres (8 feet).

The C-O sole (*Pleuronichthys coenosus*) is dark brown on its upper surface and bears a distinct black spot in the middle. It grows to 36 centimetres (14 inches) and ranges from Baja California to Alaska. Juveniles frequent shallow inshore waters; adults occur to 190 fathoms. The species gets its common name from the "C" and "O" markings on the tail. It is frequently referred to as the popeye by fishermen.

CHAPTER 11

Sea Mammals

Poking around the intertidal on some protected beach, you might have the feeling that you are being watched. You probably are. Look seaward a short distance, and you might find a smooth, dark head and a pair of equally dark eyes watching you. It's the quiet and curious harbour seal. When it has watched you long enough, it will slide beneath the surface, disappearing without a ripple.

Although harbour seals are typically solitary in the water, you may find dozens draped over the rocks together like so many dusty brown sacks. A beachwalker in search of harbour seals should look at low tide for small islands or long, narrow reefs. Seals haul out where they can see anything approaching them from either land or sea. Sea lions are not seen as often as harbour seals, but they frequently announce their presence with much hoarse barking. They are bold and can be aggressive.

If there are sea otters to be seen, the beachwalker will have no problem seeing them. Even though they are the smallest marine mammal of the Pacific coast, they spend all their time at the surface except when they are diving for food. Sleeping, eating, grooming and travelling are all done at the surface.

Beachwalkers might think that whales would be impossible to see from shore. Not so. In fact, there is a greater likelihood of seeing a whale or dolphin from the beach than there is of seeing most fish, no matter how common the fish species might be. The reason? Whales and dolphins must come up to breathe, and the tell-tale puff of water vapour as they exhale or the dorsal fin of many species gives a brief glimpse. You might only see a fraction of the animal for a fraction of a second, but it's still a thrill. Migrating grey whales can be seen from many headlands along the West Coast, and occasionally great schools of dolphins swim close enough to shore to be seen. The experience of a lifetime, though, is to share a rocky reef with a pod of killer whales, you on the intertidal, the whales in the subtidal.

ABOUT SEA MAMMALS

Marine mammals—including sea otters, sea lions, walruses, seals, whales and dolphins—are all warm-blooded, air-breathing mammals. To be air breathing on land seems quite logical. To be air breathing in water means that every marine mammal is tied to the water's surface for air. Unlike sharks and salmon, for example, which can go anywhere, a marine mammal is tied to an invisible chain. The marine mammal can only travel as deep as a round-trip lungful of air will allow; otherwise it will drown.

Warm-blooded means that marine mammals need plenty of food to generate metabolic heat, as opposed to fish, which do not generate metabolic heat independent of their environmental temperature and therefore require much less food in proportion to body size. In addition, marine mammals must find a way to keep warm and conserve heat. Think back to the harbour seal hanging quietly in the water watching you from shore. Now put yourself in the seal's place, stripped naked and sitting or swimming for five or six hours. Even if you had left your clothes on, it wouldn't matter because wet clothes are no protection in cold water. Water draws off body heat about eight times faster than air of a similar temperature. Your body would soon lose heat faster than it could produce it, your body temperature would start to drop (hypothermia), and, well, that would be the end.

So how do marine mammals stay warm? Size is one factor. No marine mammal is very small. Thus, the ratio of body mass to body surface is fairly high. The body mass produces heat; the body surface loses it. Marine mammals further reduce the surface area from which heat can be lost by having short limbs, small ears (if any) and no tails to speak of. This also explains why marine mammals have one large offspring rather than a litter of smaller ones. Looking at the marine mammals from the least specialized and most recently evolved sea otter through the seals and onto the whales, the trend is to replace fur with fat as a thermal insulator. Fat, or blubber, keeps the animals warm, helps keep them buoyant since fat is lighter than water, and provides stored energy in an environment where food cannot be cached and food abundance may fluctuate. Only the sea otter uses fur, not fat, to keep warm.

The river otter (*Lutra canadensis*) is very much at home in the coastal marine environment and the offshore archipelagos. Consequently, it is frequently mistaken for its somewhat longer and more heavy bodied cousin, the sea otter. The river otter is not a true marine mammal like the sea otter and must return to land to sleep, mate and bear and rear its young. Front and hind feet are quite different from those of the sea otter.

SEA OTTERS

Such is the beauty of the animal and especially of its skin that this otter is alone and without equal, for in the amazing beauty and softness of its fur it surpasses all other creatures of the vast ocean.

> —G. W. Steller, a German naturalist who accompanied Bering on his expedition to Alaska, 1751

In 1741, an expedition headed by Vitus Bering, a Danish explorer in the service of Russia, discovered the home of the sea otter in the Aleutian Islands. This newfound knowledge spurred a relentless hunt, which was to last 170 years, in response to the Chinese and Russian fur markets. In a continuing search for the coveted fur, the Siberians plundered island after island, slaughtering the sea otter, until they reached the North American continent.

By the end of the eighteenth century, otter herds in the western and central Aleutians had greatly diminished, causing these interesting creatures to change their way of life. Only the most violent storms could drive them to the beaches where they had once slept and given birth. They now spent their whole lives in the sea.

In 1778, Captain James Cook's crew traded trinkets for sea otter pelts from the Nootka people. When the skins reached China, they commanded such high prices that the English decided to enter the sea otter trade, breaking the Russian monopoly. Adventurers from European countries and, later, Americans were quick to join the highly profitable enterprise, leading to the near-extinction of the sea otter.

Finally, in 1911, the United States, Great Britain, Russia and Japan signed a treaty making it illegal to kill a sea otter or to possess a pelt. This treaty was primarily for the protection of the fur seal; the sea otter was secondarily included, though it was thought to be extinct. Fortunately, some animals had survived in a few secluded bays in the Aleutians, the Kuril Islands, Alaska, British Columbia and California, and very slowly their numbers increased (except in British Columbia, where they probably died out).

The sea otter *(Enhydra lutris)* is the smallest and least specialized of the marine mammals, a group that makes its living exclusively from the sea. The sea otter is not

Sea otter (*Enhydra lutris*). When not travelling, a sea otter typically spends its time at the surface on its back. Hind limbs are large flippers for swimming and diving. Front limbs are highly sensitive paws used when foraging to feel for snails, crabs, urchins and other animals hiding under rocks and in crevices.

related to whales, seals and sea lions but is the largest member of the family Mustelidae, which includes skunks, weasels, badgers, minks and river otters. Unlike its cousins, however, the sea otter does not possess anal scent glands.

Because they are both often seen swimming in coastal waters, sea otters are frequently confused with river otters *(Lutra canadensis)*. Both species are approximately the same length, 127 centimetres (50 inches), but in proportion to its length the sea otter is a heavier-bodied animal. Its tail is shorter and flattened, its hind feet are webbed, its cheek bristles are very long and mustache-like, and it has the habit of swimming on its back. When on land, sea otters are rather clumsy, whereas river otters are agile and have land-oriented limbs.

Although river otters are seen swimming in coastal waters, they are seldom found far from a source of fresh water and habitually "den-up" on land at night and to have their young. In contrast, sea otters spend the majority of their time in the water, except in remote northern regions, where they may occasionally come up on land to sleep. Sea otters do not prepare a land-based den in which to give birth and care for young.

The sea otter has little blubber, or fat layer, for insulation. How, then, does this creature maintain its body warmth in the chill waters it inhabits? To combat the cold, the sea otter has developed a pelt dense enough to provide a warm, dry blanket against its skin. The thick, fine brown fur of the sea otter is twice as dense (100,800 hairs to the square centimetre [650,000 hairs to the square inch]) as the next most densely furred mammal, the fur seal. This fine hair traps a layer of air next to the skin, providing warmth and buoyancy and allowing the sea otter to float easily at the water's surface. If the fur becomes soiled, the waterproofing and insulating quality of the outer fur is lost, the under fur becomes wet, and the animal quickly dies of cold and exposure. As a result, sea otters spend a great deal of time grooming their fur to keep it clean. This involves vigorous scrubbing of the entire body with their hand-like forepaws while blowing air into the fur. Rolling and somersaulting during the grooming flushes away debris and smooths the outer guard hairs. Because the sea otter is dependent on the condition of its fur for survival, this animal is extremely sensitive to water pollution.

Ironically, this luxuriant fur was the reason for its near-extinction a hundred years ago.

Even with its magnificent pelt, the sea otter has a high metabolic rate to help keep it warm. Maintaining this rate requires an enormous amount of food, and each individual consumes 20 to 25 per cent of its body weight each day. Thus, an average male of 30 kilograms (66 pounds) eats 7.25 kilograms (16 pounds) of food daily. Sea otters are constantly eating, subsisting on snails, clams, sea urchins, crabs, worms, shrimp, tunicates and fish. To gather these items, the sea otter makes short dives of 15.25 to 30.5 metres (50 to 100 feet), collecting food in pouchlike flaps of loose skin between the forelegs and body. Returning to the surface with its bounty, the otter rolls on its back, using its chest as a kind of lunch counter. When the meal is finished, vigorous scrubbing and rolling provide the cleanup.

Some sea otters show a most interesting behaviour pattern when feeding. Floating on its back, an individual may place a stone (or clam shell) on its chest and then proceed to pound a clam or sea urchin against the stone. Behaviourists do not consider this an intelligent ability to anticipate a need for a tool but merely an instinctive response.

Mating occurs in water, and one offspring is born every other year after a gestation period of approximately eight months. The 1.75- to 2.25-kilogram (4- to 5-pound) pup is usually born in the water and is fully developed but helpless. Female sea otters do not have a den for their young, so the mother cradles the pup on her chest as she swims on her back. Observations of wild sea otters indicate not only that there is a long period of dependency (up to one year) but also that the female otter is unusually solicitous and protective of her young. She will swim on her back, holding the pup on her chest, where it nurses at her abdominal nipples. If danger threatens, she takes the pup in her forepaws and dives underwater. She continually licks, combs and washes the young animal, leaving it only when she is getting food. At such times, the pup is left floating on its back while she dives. On land, the mother may rest on her back with the pup on her chest. If she moves around, she holds onto the pup with her teeth, letting it dangle like a limp bag.

Diurnal in their habits, sea otters congregate in

Mother sea otter cradles her newborn on her chest. A mother's chest is a floating island for a pup where it will sleep, eat and play until it is a few months old and simply too big to sleep on Mom anymore.

groups of about thirty animals, known as rafts. They segregate into "male" and "female" areas, in the water and on land. Although most activities take place in the water (eating, grooming, dozing, mating and playing), sea otters do occasionally haul out on land to dry their fur and sleep. When dozing in the water, otters are known to wrap themselves in kelp to keep from drifting.

Sea otters make a variety of sounds: they scream when distressed, coo when mating or fondling young, whistle when frustrated, snarl when captured, grunt when eating and bark when cornered. Where sea otters are accustomed to people and human activity, they are relaxed and easily observed from shore. Sea otters are nonaggressive if left alone and seem quite content to mind their own business.

Natural repopulation of areas where sea otters once flourished has been very slow. Wide expanses of ocean where the animal can obtain little food or shelter hinder migration to new territories. Thus, the natural ecological zone of the sea otter is a narrow one close to a coastline or island sheltered from wind and storm waves, with an abundant invertebrate fauna for food. While many such areas exist, it is almost impossible for the sea otter to reach them without help. To facilitate repopulation of certain areas, various government agencies have tried transplanting sea otters. A transplant involves capturing animals in an established area and transporting them to another area that once had otters and is considered suitable. Attempts have been made to repopulate the outer coasts of Oregon, Washington, British Columbia and southeastern Alaska with sea otters from the more northerly Alaska populations.

Between 1965 and 1972, a total of 708 sea otters were captured in Alaska for translocation to southeastern Alaska, the Pribilof Islands, the west coast of Vancouver Island in British Columbia, the Washington coast and the Oregon coast. Sea otters are now thriving in southeast Alaska, and a 1990 census revealed 600 sea otters in British Columbia. Pups have been seen in Washington, suggesting that the animals there are stable. However, it does not appear that the Oregon animals have survived.

Today one can only imagine what the large aggregations of sea otters looked like to the explorers as they travelled up the North Pacific coast. Interwoven with

Sea otters feed on a great variety of invertebrates and some fish. Crab shells, snails and sea urchins as shown here are opened up with large canine teeth and then chewed with large molars.

the search for the Northwest Passage and discovery of the North Pacific, the sea otter and its history are an important chapter in our natural heritage.

SEALS AND SEA LIONS

Coastal peoples have long known and used the sea dogs, or seals. We know this from drawings of these animals found on pieces of reindeer antler dated from Paleolithic times, from references to seals in works of the classical Greek poets and writers, and from many ancient beliefs. For example, a seal flipper under the pillow was considered to be a cure for insomnia, and sealskin garments were thought to protect the wearer from being struck by lightning. Today seal liver remains a source of vitamin A for many northern people in a land where fresh vegetables are not available.

The seals, which have replaced walking limbs with flippers that are more efficient for swimming, are pinnipeds, from the Latin *pinna*, "feather," and *pedes*, "feet." Pinnipeds have been divided into three groups; the walrus (Odobenidae), the fur seal and sea lion (Otariidae), and the true, or earless, seal (Phocidae). In total, forty-seven kinds (thirty-one species, sixteen subspecies) make up a world population of perhaps fifteen to twenty-five million individuals. These are found in the Arctic and Antarctic oceans and adjacent cooler waters. Walruses occur primarily on the circumpolar arctic coasts and thus are not discussed here. Some seals and sea lions have extended their ranges to more temperate or tropical areas, generally following quite closely the cold ocean currents to such places as Hawaii, Australia, New Zealand and the Galapagos Islands.

Like the sea otter and the whales, the pinnipeds are true marine mammals and derive their sole living from the sea. Seals seek their food at depths to 600 metres (1969 feet) and are able to remain under water for five to twenty minutes. As an adaptation for capturing prey, the seals and sea lions have very large eyes, which are capable of accommodating to the extremely low light levels in the darker depths. At least one pinniped, the California sea lion, is known to possess a degree of echolocation, an obvious advantage in deep-water fishing. During a dive, the metabolic rate falls, the heart beat slows to as low as one-tenth the normal rate, body temperature drops, and peripheral vasoconstriction occurs, meaning that normal blood flow is greatly reduced to all but the essential organs, such as the heart and brain. These physiological changes conserve vital oxygen during the dive when the animal is without air. So well adapted are the pinnipeds to their life in the sea that they retain only one link with the land of their progenitors; they must return to land for the birth of their young.

The fur seal and the sea lion are distinguished from the true seal by their small ear pinnae and very long, hairless flippers, which can support their bodies upright on land. The hind flippers can be rotated forward, enabling the fur seals and the sea lions to move at a fast gallop. The well-known circus seal, performing feats of balance and precision and applauding its own performance with long front flippers is, in fact, a California sea lion. The true seal or earless seal is a plump, fusiform (tapering) animal with short, furred flippers and could not possibly manage the feats of a sea lion. It has a round, smooth head with no external ears—hence the name "earless seal."

An earless seal, the harbour seal (*Phoca vitulina*).

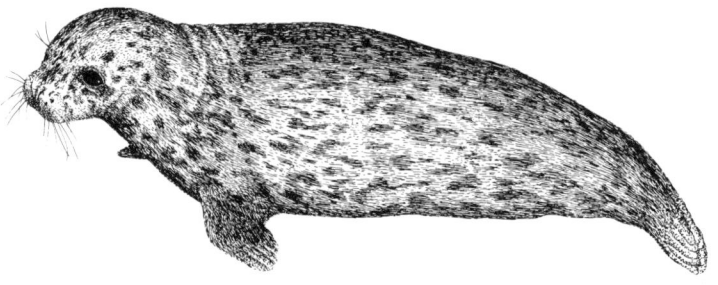

The northern fur seal *(Callorhinus ursinus)* is an occasional visitor to the West Coast. It is the most truly oceanic of the North Pacific seals, ranging across the subarctic waters of the North Pacific Ocean, Bering and Okhotsk seas, and into the Sea of Japan. It rarely comes ashore except on its home islands in the Pribilofs during the breeding season, from May through October. After the breeding season, females and young depart from the north for the warmer waters of Japan and California. It is during their passage south in the fall and north in the spring that they may be seen off Oregon, Washington and British Columbia. The males remain in the northern waters of the Bering Sea and the Gulf of Alaska. Both sexes spend up to nine months at sea, the females and young traveling 9600 to 12 600 kilometres (5965 to 7829 miles) during a single migration.

Called fur seal because of its very dense, dark fur, which has 46 500 hairs to the square centimetre (over 300,000 hairs to the square inch), this animal was once the object of massive commercial exploitation. It is estimated that 2.5 million skins were taken between 1786 and 1867, during the Russian ownership of the Pribilof Islands. Large numbers were again taken when these islands became the property of the United States in 1867. The North Pacific Fur Seal Convention of 1911 now prohibits sealing at sea except by aboriginal peoples using primitive methods. The taking of skins on land continues, but it is regulated on a sustained-yield basis by member countries of the convention: Canada, Japan, the former Soviet Union and the United States of America.

Just as true seals and eared seals differ physically, so they exhibit markedly different breeding behaviour. True seals, like the harbour seal, are promiscuous, whereas eared seals, like the Steller's sea lion *(Eumetopias jubata)* of the West Coast, have what is known as a "harem" system. This breeding behaviour results in a distinct, annual, social and biological ritual. The centre of activity is the rookery; for the Steller's sea lion, the rookery is the exposed rocky islands from California's Channel Islands northwest to the Gulf of Alaska and through the Aleutian and Pribilof islands.

The phenomenon begins with the arrival of breeding males in early spring. These are enormous buff-coloured animals, largest of the eared seals, up to 3 metres

Steller's sea lion (*Eumetopias jubata*). During the breeding season, large male Steller's sea lions establish territories on the rookery. Notice the large males in the centre of the photograph. All females are welcome on the breeding bull's territory; however, young males are not. They wait out the breeding season on a bachelor island.

An eared seal, the northern fur seal (*Callorhinus ursinus*).

(10 feet) in length and weighing in excess of a metric ton (or ton). The battle for territories begins as the largest and strongest, usually those of about twelve years of age, establish areas, or territories, of their own. A territory is approximately 55 metres square (592 feet square), depending on the physical characteristics of the land. Only those bulls that gain a territory will have access to the females when they arrive on the rookery. The younger males, while sexually mature, may be unable to compete for a territory and must retreat to another area known as a bachelor island. Much of the confrontation that occurrs between males is bluff, noise and posture. However, if one male puts any part of his body into another's territory, the defender is bound to attack his opponent with a vicious lunge at the neck, using his powerful teeth. The great neck mane and blubber of the huge males is of obvious survival value in battle.

Unlike harbour seals, sea lions are very bold and noisy. Harbour seals do not bark; sea lions do.

In June and July the females arrive, establishing themselves in whatever location they find most attractive. They are much smaller than the males, about 228 centimetres (7½ feet) in length and weighing up to 454 kilograms (1000 pounds). This difference in size is referred to as sexual dimorphism.

It was previously thought that the harem bull, or a male having a territory, chose and fought over specific females. It is now known that he simply provides real estate, so to speak, for the females, who move freely between territories, often pupping in one, mating in another and nursing in yet another.

Within a few hours or days, the newly arrived females will give birth to a single pup, the offspring of the previous year's mating. After a two-week period of frequent nursing, she will come into estrus, or heat, and be mated for the following year by the bull whose territory she is in.

But even residence by females is no guarantee of success for the battle-worn males. One intrepid biologist, studying the breeding behaviour of Steller's sea lions, had an interesting story to tell. Before the arrival of the males on the rookery, an observation blind was constructed on the windswept island. True to nature, the bulls arrived and began establishing territories. One large, but younger, bull managed to establish a very small territory, hardly much larger than himself, near

the water's edge. For this, the young gallant was prepared to fight to the death. On schedule the females
arrived, and not one took up residence on the small
rock by the sea. However, as the females began giving
birth and nursing, they soon established a routine of
leaving the larger territories to feed at sea, returning
between trips to suckle their young. Many females had
to pass the seaside rock occupied by the lone bull, and
many were eventually serviced by him as they came and
went. The bull with the most females on his territory
ended up with only the noise and congestion of mothers and pups without the benefit of any lovemaking to
make it all worthwhile.

Female Steller's sea lions are, at best, indifferent
mothers, doing little else but provide their young with
very rich milk. Seal milk is 42 per cent fat, whereas human mothers' milk is only 2 to 3 per cent fat. The richness of the milk guarantees a rapid growth rate and a
fast accumulation of a blubber layer; this layer adds
buoyancy, insulates against cold, provides a streamlined
shape that moves easily through water and serves as an
energy reserve. The growth rate of all seals is enormous
and, for some, phenomenal. For example, the southern elephant seal which is 45.5 kilograms (100 pounds)
at birth, quadruples its weight in three weeks; harbour
seal pups double their weight during the same period.

California sea lions (*Zalophus californianus*) are not as large as Steller's sea lions and may vary in colour from tan to russet to nearly black. Like all eared seals, sea lions have long front and rear flippers, can "stand" on their front flippers and have small external ear pinnae.

The young Steller's sea lion learns to swim on its own
at about three weeks of age, in the now-fetid tide pools
on the rookery. Sea lions are not concerned with sani-
tation, and, over the weeks on the rookery, wastes accu-
mulate to such an extent it is perhaps fortunate that the
sea lions have a poor sense of smell. At the age of three
months, those pups that have survived the rigours of
rookery life are ready to go to sea and learn to catch
fish, though they will continue nursing until nine
months of age or longer.

By August, the function of the males on the rookery
is complete, and they depart. Most have not eaten during the three-month season but have lived off their blubber and must now replenish their body stores in
preparation for winter. The females follow within a few
months, ending the complex but temporary society of
life on the rookery.

Often confused with the Steller's sea lion is the California sea lion *(Zalophus californianus)*. This species is

much like the former in physical appearance and behaviour but is smaller, darker in colour (russet to brown) and more southern in its distribution (Mexico to southern California, with a non-breeding range extending northward into British Columbia). The male California sea lion may reach 244 centimetres (8 feet) and 363 kilograms (800 pounds). The female is much smaller at 152.5 centimetres (5 feet) and 113 kilograms (249 pounds). Like the Steller's sea lion, the California sea lion engages in seasonal breeding behaviour.

The most common true seal of the West Coast is the harbour seal *(Phoca vitulina)*, found in bays and inlets from China to the Bering Sea and south to Baja California. Harbour seals grow to 114 kilograms (251 pounds) and 156 to 187 centimetres (5 to 6 feet) in length. They swim using a sculling motion of the hind flippers, the short tail tucked neatly between them. On the water they are recognized by the manner in which they approach the surface for air, rising from directly beneath the surface and hanging vertically in the water while surveying their surroundings. Because they are unable to rotate their hind flippers forward, they move on land in a caterpillar-like motion, using the front flippers to pull themselves forward and dragging the hind flippers or holding them aloft. Lacking swift mobility on land, true seals, like the harbour seal, seldom haul out more than a few yards from shore and easy escape from land predators. Although they are able to sleep comfortably in the water, these seals generally haul out to sleep and to enjoy the benefits of dryness and sunshine for a good part of the day.

Harbour seals generally sleep on their sides, and when they sense something approaching, they will lift their heads and hind flippers at the same time, forming a chubby crescent. If you want to watch them out of water, proceed very slowly when you see heads raised. As soon as any seal in the group starts to roll onto its belly in preparation for a slide into the water, you should stop because chances are your next step will send all the seals into the sea.

Harbour seals mate about September, usually in the water, though sometimes on land, especially on sandbars. The young are born in spring or early summer on a secluded sandbar or beach. Pups are 8 to 11.3 kilograms (18 to 25 pounds) at birth, fully furred and able

The harbour seal (*Phoca vitulina*) is a common, inshore and year-round resident of both coasts of North America. These seals are often seen hauled out on sandbars and secluded beaches. A single 10-kilogram (20-pound) pup is born in June or July. Unlike the Steller's sea lion pup, which may nurse for up to a year, the young harbour seal is weaned by six weeks of age.

Harbour seals are distinguished by short front and hind flippers and no external ear pinna—hence the name "earless seal."

to swim. Pups born in the southern areas of their range bear the same brown, tan and black mottled fur pattern as the adults, whereas those born in the northern areas may retain for a short period a white fetal coat, an adaptation to early life in a land of snow and ice. Harbour seals are gentle and devoted mothers, playing swimming games with their young and responding quickly to their human-sounding cry. A mother may leave her young momentarily to feed, or if disturbed, but soon returns to retrieve her pup. A well-meaning person should make absolutely certain, by retreating to a distance and watching for the return of the mother, that a pup is truly abandoned before taking steps to take and care for it.

The nursing period for harbour seals is short—four to six weeks—by which time the pup has a good set of sharp teeth and is weaned to an adult diet of fish, mollusks and crustaceans.

A relative newcomer to West Coast pinnipeds is the elephant seal *(Mirounga angustirostris)*—an enormous animal. What this seal lacks in beauty, it more than makes up for in size. The male weighs up to 3.25 metric tons (3 tons) and exceeds 5.5 metres (18 feet) in length. Females are smaller, at 1 metric ton (or ton), and measure 3 to 3.5 metres (10 to 12 feet). Like the land-living elephants, they are relatively hairless. The males of this species develop a strange proboscis, reminiscent of the trunk of the land elephant. This bulbous snout is inflated during courtship battles, and when relaxed, it hangs drooping over the mouth. This mammoth animal generally breeds in the warmer climes of southern California and Mexico, and the bulls are infrequently seen in British Columbia waters. Unlike the other members of the true seal family (except for the grey seal of the Atlantic), the elephant seal has a kind of harem system, breeding in January and February on rookeries in southern California and Mexico. Again, unlike other phocids, the young cannot swim at birth, taking to the water at two months of age.

Like the northern fur seal, the elephant seal was the object of commercial sealing to such an extent that by 1900 it was thought to be extinct. The animal was taken not for its fur but for its blubber; an average 3.5-metre (12-foot) bull yields about 410 litres (108 gallons) of oil. The sealing that began in 1818 was no longer profitable by 1860, so extensive was the kill. In 1922, the

Young male elephant seals (*Mirounga angustirostris*) posturing on the beach in California. This species is by far the largest earless seal in the world. Males may grow to 5.5 metres (18 feet) and weigh up to 2720 kilograms (6000 pounds).

Mexican government granted the elephant seal absolute protection.

In the United States protection has been granted to all marine mammals, including seals, under the Marine Mammal Protection Act of 1972. Basically, this act states that it is illegal to take, hold, harass or kill for any purpose any marine mammal, or tissues thereof, without a permit. In Canada, it is illegal to take, catch, kill, molest, disturb or be in possession of an elephant seal, sea lion or seal without a permit, except where a seal is causing damage to fish, nets or gear associated with those who earn their living by fishing. This protection is also extended to sea otters and killer whales.

Legislation can provide a seal or sea lion some degree of protection from human activity but not from their natural enemies. The greatest single killer of pinnipeds is infestation by a small parasite, the nematode. This creature is responsible for at least 22 per cent of fur seal pup mortality. Tapeworms, hookworms, nasal mites and lice are other common parasites that though they may not kill the animals directly weaken them to a degree that normal avoidance of predators is impossible. Heaviest mortalities from all causes occur in the juvenile animals, the pups of many species falling prey to leopard seals, walruses, polar bears and bald eagles. The greatest enemy of the adult seal is the killer whale in cooler waters, and large sharks in temperate and warmer seas.

WHALES AND DOLPHINS

Whales and dolphins are mammals: they are warm-blooded and breathe air. Through extreme adaptation to life in the water, from birth until death they are totally independent of the land. Such aquatic mammals are called cetaceans. Small cetaceans are called dolphins or porpoises; larger ones are called whales.

Cetaceans bear small resemblance to their land-dwelling ancestors. Just why a mammal would return to the sea and to the problems of mammalian existence in the water is not known. Perhaps it was in response to the need for more space or a new food source, or to escape predation.

The cetaceans are divided into two groups according to their very different feeding patterns. Most of the smaller whales, dolphins and porpoises are "toothed whales," or odontocetes. (There is, however, one large toothed whale, the sperm whale.) As the name implies, these cetaceans have teeth. They feed on fish and invertebrates such as squid and crustaceans. The diet of the killer whale is more varied and includes sea birds and marine mammals.

Of the larger whales, all but the sperm whale belong to the group known as the baleen whales, or mysticetes. In this group, teeth have been replaced with large structures called baleen. These plates of horny material hang like vertical venetian blinds from the upper jaw. They number 200 to 400 per side and are frayed into bristles on their inner edges. These elaborate structures act as sieves to capture plankton or krill.

Plankton, found in the upper layers of the ocean, ranges from microscopic larval forms to small shrimps a few centimetres in size. Plankton serves as a primary food source for the baleen whales, as well as a host of other marine mammals.

To trap its food, a baleen whale swims through the plankton-rich sea, mouth open, engulfing plankton and sea water by the ton. By shutting its cavernous mouth and pressing its tongue against the back of the baleen bristles, it forces the water out of its mouth, trapping the plankton on a mat of overlapping baleen. By using a primary food source, the baleen whale is able to secure the enormous amounts of energy required to sustain its huge body.

A baleen whale—the blue whale—is the largest animal ever to have lived on Earth, larger by far than any dinosaur. This leviathan has been measured up to 30.5 metres (100 feet) in length with a corresponding weight of over 110 metric tons (100 tons). The great size of this and other whales has been possible because they are suspended in water and do not have to support their weight on limbs against the pull of gravity, as do land animals.

Whereas land mammals are able to breathe equally well through both nose and mouth, cetaceans have access to air only by way of a modified nostril, or blowhole, on the top of the head. A series of muscles surrounds this blowhole, sealing its entrance when the animal submerges.

One might well ask how a whale (or dolphin), an air breather, is able to swim under water with its mouth open and never run the risk of flooding its lungs. The answer is the goosebeak larnyx, which leads air from the blowhole across the mouth cavity into the trachea.

Cetaceans do not breathe in the rhythmic, involuntary fashion of land mammals but inhale and exhale according to conscious effort, quickly and with force, literally blowing air from the lungs.

Cetaceans do not spout water. The visible spout, the size and shape of which is unique to many species, results from condensation of warm vapour entering the air from the lungs as the animal exhales. The spout effect is a result of rapid and forced exhalation of water vapour and a small amount of water present in the depression around the blowhole.

Unlike the baleen whales, which feed near the water's surface, some toothed whales and dolphins dive to great depths in search of food. For example, the sperm whale is known to dive regularly down 457 metres (1500 feet) in search of prey. Consider that once this whale has left the air-water interface, it must make a 1-kilometre (⅔-mile) round trip without breathing.

Because cetaceans have a larger blood volume than land mammals of comparable size and weight, they have a greatly increased capacity to store oxygen in their blood and muscle tissue. In addition, they have a more efficient system of supplying oxygen to the blood: each breath provides an 80 to 90 per cent renewal of air in the lungs, compared with a renewal of only 10 to 20 per cent in most land mammals. Yet another important adaptation for prolonged breath holding in cetaceans is their resistance to the metabolic by-product carbon dioxide.

Contrary to what is commonly thought, it is the build-up of carbon dioxide in the tissues, not lack of oxygen, that triggers the involuntary breathing response of most mammals. Because cetaceans have a high tolerance to carbon dioxide, they are able to remain submerged for longer periods without being overcome by the need to breathe. Even so, cetaceans must "pant" at the surface to replenish their oxygen supply after a long dive, in much the same way as a sprinter pants at the end of a race.

At first glance, whales and dolphins appear remarkably fishlike. In truth, they are as far removed from fish as are human beings. Although the cetacean body has become exceedingly streamlined, it is based on mammalian structure. Front limbs have become modified as paddle-shaped flippers, the bones of which are similar to those of jointed limbs and digits. Hind limbs have been lost, the bones of the pelvic girdle now serving only as anchors for the reproductive organs. Tail flukes, which provide the main propulsive thrust of these animals, bear no anatomical connection to hind limbs but are a separate and distinct development. Tail flukes contain no bone, owing their firm yet flexible shape to underlying fibrous and elastic tissue. In whales and dolphins, tail flukes are always horizontal because the animals must be able to swim up towards the water surface for air. Fish have vertical tail fins because they only need forward thrust, not forward and up.

To provide a fluid, fusiform shape offering the least resistance as the animal moves through water, the internal organs, skeleton and muscles are enclosed in a thick layer of blubber, smoothing and rounding out physical irregularities. The skin is modified as well, free of sweat glands, oil glands and hair (except for a few facial bristles in some species), and feels much like smooth, wet rubber when touched.

Possibly the most fascinating aspect of whales and dolphins is not their extraordinary physical adaptations and capabilities but their nonphysical ones. Theirs is a world perceived largely through sound. Sound and hearing are to the whales and dolphins what vision is to most land mammals.

For years scientists have been intrigued by the ability of cetaceans to explore their environment and the objects in it through the use of echolocation. By directing sounds produced in the head region towards an object and receiving the sound waves that bounce off the object, a cetacean can make very fine discriminations as to size, density, distance and so on. The sound waves are received a pulses though the lower jaw and transmitted to the inner ear. Because sound waves are waterborne, cetaceans have been able to discard the structures that land mammals developed to gather airborne sounds, namely, the external ear.

This system of sensing the environment is an enormous advantage in orienting, navigating and capturing prey in dark or turbid waters. In essence, it is a means of scanning by sound for the same information we per-

Killer whales (*Orcinus orca*) generally travel in family groups known as pods. Adult males have tall, straight dorsal fins; females and immature males have shorter, crescent-shaped dorsal fins.

Killer whales occur throughout the world in temperate seas. They often hunt in packs, using their forty-four to fifty sharp, conical teeth to tear flesh from seals, sea lions and even other whales. Killer whales resident along the West Coast tend to be fish eaters.

ceive by vision. This is not to say that cetaceans have poor sight. Researchers found that the visual acuity of the killer whale under water was equal to that of a cat on land.

In whales and dolphins, sound communicates much of what other animals communicate through scent, posture and expression. Cetaceans lack any sense of smell. You will never see a whale curl its lips in a snarl or wag its tail. A female will not emit a distinctive scent, indicating a readiness to mate. In cetaceans these important messages are relayed by sound.

Because whales and dolphins are extremely difficult to observe in their natural state, little is known about how they communicate and behave in the wild. Some species exist, for the most part, as solitary animals. How do they locate a mate? Other species occur in pods numbering from a few animals to hundreds of individuals. How are they organized?

Research on cetacean physiology and behaviour is still in its infancy. The animals are difficult to study in their wild environment for two reasons: they are large and often fast moving, and they are only visible at the surface when they come up to breathe. Breathing is 5 to 10 per cent of their activity. Most of what cetaceans are doing is hidden from view when they are under water. As a consequence, much of what has been learned is based on work done with small cetaceans at zoos or aquariums or is the result of post-mortem studies on dead, stranded whales or those killed during whaling operations. Radio telemetry and satellite tracking from transmitters attached to the whales is providing much new information on animal movements and diving behaviour. It is tempting to generalize when information is scant, yet caution must be used in extrapolating what has been learned from a few species and applying it to all cetaceans.

Although many details of cetacean behaviour remain a mystery, it is known that all cetaceans pair and ultimately mate. Brief mating occurs in the water. The pregnant female carries her unborn calf for eleven to sixteen months, depending on the species. As parturition approaches, she probably seeks out a sheltered area and gives birth to a single calf, one-quarter to one-third her own length. The calf is born head or tail first and swims immediately towards the surface for its first breath. For up to a year and perhaps beyond, the calf will be nursed on rich mother's milk from two mammary teats enclosed in slits located on either side of the genital opening.

The mother-young bond is a strong one. The whalers, like the sea-otter hunters, capitalized on this strong attachment, capturing or killing the slower and weaker young in order to secure the adult, knowing that a mother would not leave her calf. Even in an undisturbed state, all young are not destined to reach adulthood. Disease, predators and natural disasters prevent overpopulation.

Mother killer whale "spy hopping" to survey danger or prey above the water. Experiments indicate killer whale eyesight equals that of a cat in air. White patches of the adult are lemon-yellow in the young killer whale.

Because cetaceans breed slowly, producing only one offspring every two or three years, many were unable to withstand the pressures of modern commercial whaling. As a result of hunting by high-speed whaling vessels equipped with explosive weapons in pursuit of animals that cannot run and cannot hide, there are few whales of some species now, whereas formerly there were many. A five year pause in commercial whaling and the establishment of some whale sanctuaries has provided relief of some species that were previously hunted. Whether or not some, or all, of the traditional whaling nations will resume whaling in the future is not known at this time.

The relatively slow swimming speed of the California grey whale (4 to 5 knots cruising speed) and predictable seasonal migration routes were well known to the

False killer whale (*Pseudorca crassidens*) is so named because it looks similar to the killer whale. It occurs throughout the world and may come close to shore on the West Coast.

whalers of the early 1800s. As a result, grey whales were severely depleted by the 1830s. Under protection by international agreement since 1838, the grey whale has increased, and great numbers of these fascinating animals can be seen passing our shores on their 12 874-kilometre (8000-mile) migration.

In the fall, the whales can been seen moving south off the West Coast en route to winter breeding grounds in the lagoons of southern California and Mexico. The springtime finds this baleen whale travelling north by the same route to summer feeding grounds in the Gulf of Alaska and the Bering Sea, where it feeds on amphipods scooped off muddy bottoms. It is now thought by some biologists that a population of grey whales, most likely a nonbreeding group, remains year round in the region of Pacific Rim National Park.

The grey whale is slender, with a blunt head and dorsal hump, followed by several lesser humps. Its colour is mottled grey or blackish, and it has a mature length of 13.5 metres (44 feet).

In sharp contrast to the slow, peaceful and rather drab-coloured grey whale is the impressive strength, speed and colour pattern of the killer whale *(Orcinus orca)*. The distinctive white markings on a sleek and solid black background indicate that this species does not need to camouflage its presence from enemies, and in fact it has no enemies other than humans.

Killer whales are so called because of their predatory nature. They hunt in packs of three to forty, feeding on fish, birds, seals, sea lions, porpoises and even larger whales. Food is not chewed but is torn into chunks by the powerful jaws and sharp teeth.

Killer whales are distributed throughout the world, with their largest concentration thought to be on the West Coast. They have no known definite migration and appear to remain from one season to the next within a loosely established home range. Alert Bay, off the central east coast of Vancouver Island, has perhaps the largest resident killer whale population. Here it is possible to see up to forty animals in a small area. The reason for this concentration is not certain, but it is thought to be connected with the fact that the region is particularly rich in salmon. Haida folklore and art of the Alert Bay region strongly reflect the presence of the killer whale and attest to its presence there for many years.

Killer whales are seen regularly off Alaska, British Columbia and Washington. The presence of killer whales in the wild is unmistakable, heralded by what can only be described as powerful exuberance. Leaping and blowing, with large black dorsal fins cutting through the seas, they are a thrilling sight. Killer whales are seldom seen alone, so chances are you will see a group or pod. If the whales are travelling, your observation will be over far too soon. Black fins appear and grow taller and taller in the water as the whale advances and rises to the water's surface for air. It exhales, sucks in and rolls downward in one smooth and powerful motion. The black fin shrinks and disappears behind the whale. If the whales are feeding, their time with you will be longer, as individual whales turn this way and that, some diving, some circling. The animals spread out, then come together, disappear and reappear. They are so close, so completely at home, so unafraid and so big. Sharing a beach with killer whales is a spine-tingling experience, the first time and every time. Although killer whales are not known to prey on humans, they are predators of warm-blooded mammals and should be respected as such.

Male killer whales grow to 9 metres (29½ feet) in length, whereas females remain somewhat smaller at 6 to 6.75 metres (20 to 22 feet). The size and shape of the dorsal fin differs in the sexes, the females having a sickle-shaped fin that is shorter in proportion to body size than that of the male. The dorsal fin in the male killer whale is triangular and can extend to 1.75 metres (6 feet) in height in a mature bull.

Killer whales are one of the best-studied cetaceans in the wild and are a prominent feature of West Coast marine fauna, if not in numbers at least in public awareness. Killer whales are easily identified in the wild, have a spectacular reputation and have been popularized by aquariums and oceanariums. There is a liberal blending of fact and fiction about whales in general and killer whales in particular. As more and better research is carried out, it is hoped that whales will be accepted for what they are—magnificently adapted animals functioning superbly in an environment that is alien to humans.

The cost of conducting field research is a very real factor in determining which animals are studied and which are not. Because the inshore pods (also known as resident pods) spend summer and fall near shore, killer whales are accessible to researchers in small boats with inexpensive equipment (in contrast to studying offshore whales on the high seas). The grey saddle patches just foreward of a killer whale's dorsal fin provide researchers with a natural identification system. Thus, whales can be photographed, identified and followed from season to season, providing invaluable information on the stability of their social relationships, seasonal movements, longevity and reproductive cycles. It must be remembered, however, that what is learned about the Alaska, British Columbia and Washington inshore whales is a snapshot of a few hundred killer whales living in one of the richest food areas available to the species and may not be applicable to killer whales in other parts of the world.

The Pacific white-sided dolphin *(Lagenorhynchus obliquidens)* ranges in the Pacific from California north across the Aleutian chain to Japan, wintering in warmer waters to the south. A big dolphin, up to 2.25 metres (7 feet) in length and 90 to 113.5 kilograms (198 to 250 pounds) in weight, it travels in large groups or pods of up to two thousand individuals. It is a beautiful dolphin, black on the back with striking light grey sides and white belly. At a distance, the Pacific white-sided dolphin may

Pacific striped dolphin (*La-genorhynchus obliquidens*). Ranging from California north and across the Aleutian chain to Japan, this species travels in very large aggregations of up to two thousand individuals. The dolphin shown here is pursuing one of its major food items, the herring. Many needle-sharp teeth grasp prey, which is not chewed but swallowed whole.

be confused with the Dall's porpoise *(Phocoenoides dalli)*, which is coloured somewhat like a killer whale—black with a large white patch on each side. Unlike Dall's porpoises, Pacific white-sided dolphins are great jumpers and leapers. Dall's porpoises do not jump, but they have greater speed and are considered the fastest-swimming of marine mammals, capable of speeds in excess of 40.25 kilometres per hour (25 miles per hour).

The terms "dolphin" and "porpoise" give rise to a great deal of confusion and argument. "Porpoise" is generally used to denote a small cetacean having a blunt nose, spade-shaped teeth and a triangular dorsal fin. "Dolphin" refers to a small cetacean having a beaklike rostrum (snout), needlelike teeth and a sickle-shaped dorsal fin. The words "dolphin" and "porpoise" are actually arbitrary terms. The cetologists (those who study cetaceans) consider "porpoise" to be the more correct for describing small cetaceans, thereby avoiding the confusion between the dolphin mammal and the dolphin fish.

The Pacific harbour porpoise *(Phocoena vomerina)* is a relatively common inshore species. It is distinguished from the Pacific white-sided dolphin and the Dall por-

Dall's porpoise are frequently misidentified as baby killer whales. They are high-speed swimmers and frequently ride and play in the bow wave of a fast-moving vessel. Adults are typically 2 metres (6½ feet) in length.

poise by having a dark grey to black upper surface with a paler belly and no white patches on the sides. This species does not school but travels in small groups of two or four animals. The Risso's dolphin *(Grampus griseus)* is seen less often on the West Coast, and although the northern right whale dolphin *(Lissodelphis borealis)* has not yet been confirmed as occurring off British Columbia, it has been recorded in Puget Sound. Other members of the toothed whale group known to occur on the West Coast are the sperm whale *(Physeter catodon)*, Baird's beaked whale *(Berardius bairdii)*, Cuvier's beaked whale *(Ziphius cavirostris)*, Stejneger's beaked whale *(Mesoplodon stejnegeri)* and Hubb's beaked whale *(Mesoplodon carlhubbsi)*.

Of the baleen whales other than the previously mentioned grey whale, the fin whale *(Balaenoptera physalus)*, sei whale *(Balaenoptera borealis)* and minke whale *(Balaenoptera acutorostrata)* are known to occur on the West Coast. The blue whale *(Balaenoptera musculus)*, the humpback whale *(Megaptera novaeangliae)* and the very rare North Pacific right whale *(Balaena glacialis)* are rarely seen, since overhunting has drastically reduced their numbers.

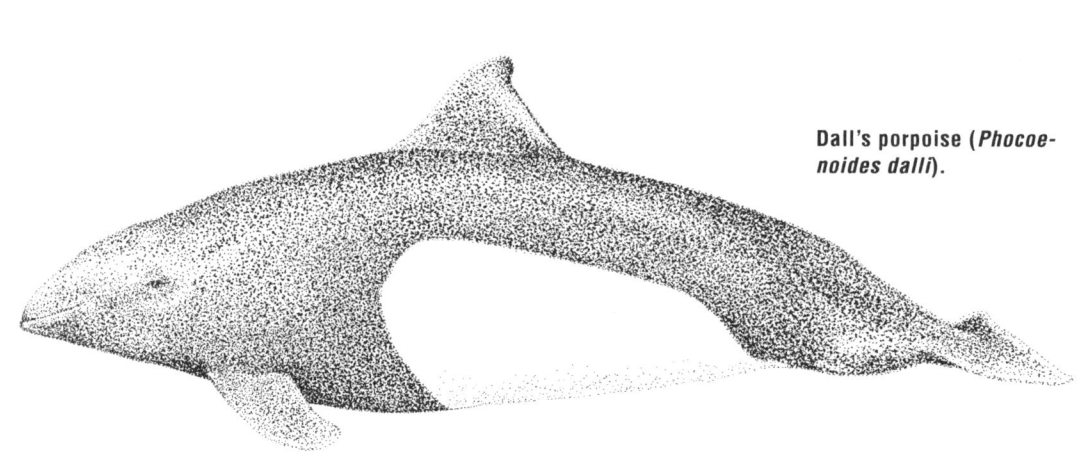

Dall's porpoise (*Phocoenoides dalli*).

FOR FURTHER READING

ABOUT THE OCEAN

Rand McNally. 1979. *The Rand McNally Atlas of the Oceans.* Chicago: Rand McNally.

INVERTEBRATES

Morris, R. H., D. P. Abbott and E. C. Haderlie. 1980. *Intertidal Invertebrates of California.* Stanford, Calif.: Stanford University Press.

Pearse, V., J. Pearse, M. Buchsbaum and R. Buchsbaum. 1987. *Living Invertebrates.* Pacific Grove, Calif.: Boxwood Press, and Palo Alto, Calif.: Blackwell.

FISH

Gotshall, D. W. 1981. *Pacific Coast Inshore Fishes.* Los Osos, Calif.: Sea Challengers, and Ventura, Calif.: Western Marine Enterprises.

Hart, J. L. 1973. Bulletin 180. *Pacific Fishes of Canada.* Ottawa: Fisheries Research Board of Canada.

Lamb, A., and P. Edgell. 1986. *Coastal Fishes of the Pacific Northwest.* Madeira Park, B.C.: Harbour Publishing.

MARINE MAMMALS

Haley, D., ed. 1986. *Marine Mammals.* 2d ed. Seattle: Pacific Search Press.

King, J. E. 1983. *Seals of the World.* London: British Museum of Natural History.

Watson, L. 1981. *Sea Guide to Whales of the World.* New York: Dutton.

INDEX

168